Multi-Objective Optimization for Urban Drainage Rehabilitation

DISSERTATION

Submitted in fulfillment of the requirements of
the Board for Doctorates of Delft University of Technology
and of the Academic Board of the UNESCO-IHE
Institute for Water Education
for the Degree of DOCTOR
to be defended in public on
Wednesday, 21 of March 2012, at 10:00 hours
in Delft, The Netherlands

by

Wilmer Jóse BARRETO CORDERO

Master of Science in Hydroinformatics, UNESCO-IHE, The Netherlands

born in Barquisimeto, Venezuela

ii

This dissertation has been approved by the supervisors:

Prof. dr. R.K. Price
Prof. dr. D.P. Solomatine

Composition of Doctoral Committee:

Chairman	Rector Magnificus Delft University of Technology
Vice-Chairman	Rector UNESCO-IHE
Prof. dr. R.K. Price	UNESCO-IHE/Delft University of Technology, supervisor
Prof. dr. D.P. Solomatine	UNESCO-IHE/Delft University of Technology, supervisor
Prof. dr. ir. N.C. van de Giesen	Delft University of Technology
Prof. dr. ir. F.H.L.R. Clemens	Delft University of Technology
Prof. dr. D.A. Savic	University of Exeter, United Kingdom
Dr. Z. Vojinovic	UNESCO-IHE
Prof. dr. ir. J.B. van Lier	UNESCO-IHE/Delft University of Technology, reserve member

CRC Press/Balkema is an imprint of the Taylor & Francis Group, an informa business

Published by:
CRC Press/Balkema
PO Box 447, 2300 AK Leiden, the Netherlands
e-mail: Pub.NL@taylorandfrancis.com
www.crcpress.com - www.taylorandfrancis.co.uk - www.ba.balkema.nl

ISBN 978-0-415-62478-7 (Taylor & Francis Group)

to my lovely mother and
to my daughter Ana Lía

Summary

Flooding in urbanized areas has become a very important issue around the world. The level of service (or performance) of urban drainage systems (UDS) degrades in time for a number of reasons: structural deterioration, siltation, new developments, climate change etc. In order to maintain an acceptable performance of UDS, early rehabilitation plans must be developed and implemented.

Cities are growing fast, and budgets for the rehabilitation of urban drainage grow at a far slower rate than budgets for urban development. In developing countries the situation is serious, little investment is done and there are smaller funds each year for rehabilitation. The allocation of such rehabilitation funds must be "optimal" in providing value for money. However this task is not easy to achieve due to the multicriteria nature of the rehabilitation process, taking into account technical, environmental and social interests. Most of the time these are conflicting, which make it a highly demanding task.

Multiobjective optimization approaches adopt appropriate tools and facilities to simplify the optimal rehabilitation of a UDS. Heuristic and Genetic Algorithms have been applied and have proven to be efficient for multiobjective problems. However, the large number of possible solutions (or scenarios) in UDS and the number of function evaluations needed by, say, evolutionary algorithms (EA) makes their application difficult for practitioners.

The present research is aimed at defining a framework to deal with multicriteria decision making for the rehabilitation of urban drainage systems, and focuses on several aspects such as the improvement of the performance of the multicriteria optimization through the inclusion of new features in the algorithms and the proper selection of performance criteria. The new framework, called a "*Multi-tier Approach*", must be suitable for use in developing countries, be scalable and be able to provide several solutions in an elapsed time that is suitable for practitioners.

A review of the state-of-the-art in urban drainage rehabilitation has been done. During a rehabilitation process several aspects have to be addressed. Issues such as the determination of performance indicators for hydraulic, structural and environment assessment have to be considered. Data availability and the identification of critical pipes and channels are also of major importance in any rehabilitation plan. Hydrological and hydrodynamic modeling plays a key role during for the hydraulic, structural and environmental assessment. Dual modeling of the above and below-ground systems is preferred in order to evaluate the surcharge consequences for the different assessments. There now exist mature computational modeling packages for 1D and 2D modeling, however, the interaction between the models produces by these packages is still a matter of research for them to become available for practitioners. Sustainable approaches, oriented to the control of runoff volumes from the beginning of the rainfall are preferable than methodologies based on conveyance. These sustainable approaches are also oriented to keep environment, social and economical values in balance. Different hydrodynamic modeling packages were reviewed in order to select the most suitable for this research, based on their advantages and disadvantages.

An approach for urban drainage rehabilitation using a multi-tier framework is developed.

This approach is innovative in that it introduces hydrodynamic modeling inside a multi-objective optimization process, using parallel computing to make it attractive for practitioner. It is modular and flexible depending on the data availability, making it suitable for the use in developing countries. It allows for the inclusion of expert knowledge at different stages of the optimization process. The framework application is simple but does not omit important features for the rehabilitation of drainage systems. The use of external tools for modeling, optimization and visualization allow for scalability which implies that the tools can grow as much as needed. A prototype framework was built and successfully tested on real problems in developing countries.

A review of the state-of-the-art for optimization in engineering problems was completed. It shows that there is no unique method for optimizing all kinds of problems, instead an appropriate methodology has to be selected taking into account the problems being optimize. Few applications in the area of urban drainage were found, and no application to real cases using hydrodynamic models inside the optimization algorithm unless they have been applied to small networks only. A special type of problem was identified called a "*highly computationally demanding*" problem, also defined in some literature as an "*expensive problem*". Multi-criteria urban drainage rehabilitation is classified as such a kind of problem where large computing power is required. An approach to face such a kind of problem is developed and tested for multi-objective algorithms using four benchmark functions. The method is based on the assumption that practitioners need only a few solutions and not a large set that are similar to each other. The new approach outperforms NSGA-II on three of the benchmarks, while NSGA-II was slightly better on the fourth benchmark function.

The proposed multi-tier framework for the rehabilitation of an urban drainage system was implemented and tested, performing a "*proof of concept*" on a small study case. Firstly, a structure for estimating the investment cost was implemented; features like pipe replacement, storage tanks and ponds, and diversion structures were included. A method for damage cost estimation was also incorporated. The cost structures are based on equations in common use and follow a unitary price analysis, including the preset worth in the cost estimation.

A prototype tool was developed and tested. A small study case was used as a "*proof of concept*". Two multi-criteria algorithms based on genetic algorithms were tested; they were NSGA-II and ε-MOEA. NSGA-II was incorporated onto a library and into an optimization tool called NSGAX. In order to compare the performance of the algorithms, four metric indicators of the multi-criteria performance were used: cardinality, time, number of functions evaluations, hyper-volume and ε-Indicator. Using these indicators it was concluded that NSGA-II outperforms ε-MOEA in general, but the differences were not substantial. Hyper-volume and ε-Indicator show convergence, and they were used as stopping criteria during the optimization process. The application of the framework to a larger network with 63 pipes showed the need for more efficient algorithms or computers to reduce computational time. In both case studies of 12 and 63 pipes, it was possible to find rational solutions that may be expected by practitioners; for instance the storage tanks were properly selected depending on their location and were also satisfactorally dimensioned. It is concluded that the developed framework is suitable for use in the rehabilitation of urban drainage systems. It shows scalability and flexibility. The use of intangible costs was also evaluated through the implementation of an objective function to minimize the anxiety of the population. The tool allows for interpretation and negotiation between stakeholders over a set of alternatives, and is not reliant in only one optimal solution.

The state of the art regarding parallelism for computer code was reviewed. Advantages and disadvantages of the different existing methods for parallel application have been studied. The theory of concurrency for multiple processors computers and cluster was reviewed. An implementation of parallel code for NSGAX was done using PVM libraries over Cygwin, a Linux OS emulator for Windows. The parallel code was included on the multi-tier optimization framework and applied to two case studies.

A small cluster composed of heterogeneous PCs with single and multi-core processors was set up. Two case studies were tested on the parallel framework: one using a small network with 12 pipes and the other for a sub-catchment in Belo Horizonte Brazil composed of 168 pipes. The results show a good saving in performance between 60% to 80% of the consumed time when the methodology is compared with the performance for a single optimization. Also, it can be concluded that the number of processors to use in the cluster has to be related to the size of the problem; if the results are analyzed using an efficiency indicator, it is better use few processors than use several of them. When the 12 pipes problem was solved using 6 processors it behaved like only 3 processors in term of speedup while for the Belo Horizonte case 6 processors behaved like 5 of them, so it was more efficient for a large problem than for a small one.

Applications of urban drainage rehabilitation with multi-objective algorithms based on population optimizers like GA is limited among practitioners. Furthermore there are less practitioners using hydrodynamic models for flood damage estimation. This is due to the methodologies being computationally demanding, encouraging practitioners to avoid them. The developed multi-tier approach was applied to two real studies case in developing countries. One case study was in Cabudare in Venezuela and the other in Cali in Colombia. It is demonstrated that a multi-tier framework is also applicable in developing countries where data is limited and simplified tools are acknowledged as being valuable for practitioners.

In the Cabudare case study, an approach using expected damage cost was applied. It allowed the discovery of three equal economic solutions over the Pareto set. A comparison with the recommended approach of the Venezuelan standard shows that these three solutions are equivalent to the determination of the complete flood damage costs for a return period of 1:10 rainfall. It also provides a set of several solutions in which there is a tradeoff between investment and damage. The NSGAX algorithm was improved to allow the inclusion of expert knowledge, through the injection of good genes into the initial population. This novel approach proved to be efficient for computationally demanding problems.

In the Cali study case the environmental variable was addressed. The multi-tier frame was coupled with SWMM 5.0 to compute flooding and water quality. A diversion was included to direct water from the network to a temporary storage for flood reduction and designed to catch the first foul flush. Despite the simplification of the storage model, the tradeoff between the investment cost and flood damage, between investment cost and water quality and the correlation between flood damages and water quality improvement can be seen.

The objective of developing and testing a framework for the rehabilitation of urban drainage networks has therefore been achieved. A prototype of the multi-tier approach was applied to several case studies in developing countries, demonstrating that it is feasible and easy to apply. Parallel computing and other methodologies make the approach attractive to practitioners, and these methodologies have been included in the framework.

Acknowledgment

Foremost, I would like to thank my supervisor, Prof. Roland K. Price, who shared with me much of his expertise and research insight, I know he had a hard time trying to motivate me, but in the end he did. I would also like to express my gratitude to Prof. Dimitri Solomatine, whose thoughtful advice often served to give me a sense of direction during my PhD studies. It is difficult to not overstate my appreciation to Dr. Zoran Vojinovich, who brought to me the idea of research in this topic; I learned much from him about modeling and practical issues on urban drainage.

Institutionally, I am deeply grateful to the University *"Centrocidental Lisandro Alvarado"*, Venezuela for the trust and support that they gave me in order to study in The Netherlands. Also I want to thank the managers of the SWITCH Project, who gave me financial support to attend conferences during this research. Thanks also to IHE for giving me the opportunity to develop this research.

I wish to thank everybody with whom I have shared experiences in life. From the people who first persuaded me to be interested in the study of hydraulics, especially those who also played a significant role in my life. Mr Nestor Mendez has been more than my lecturer; he has also been more than a friend. Arlex Sanchez not only shared his knowledge with me, providing me with the Cali case study, but he also let me enter into his beautiful home. He and his wife and children have become part of my family. Thank you too Carlos Velez, Gerald, and Leonardo, for being my "*brothers*" not only because weare from brother countries, but also because we look and behave like brothers.

Thanks also to Carlos Lopez, Kitiwet, Gabriela, Heaidi, and all my PhD fellows for being some of my very best friends. I will miss the "*Empanadas Bailables*". Also I would like to thank Natahsja, Ailen and David for their invaluable support and making me feel at home. I cannot finish without saying how grateful I am to my family: grandparents, uncles, aunts, cousins and nephews, all have given me a loving environment in which to develop. Particular thanks, of course, to Altidoro and Guiomar my brothers and best mates. Lastly,and most importantly, I wish to thank my parents. They have always supported and encouraged me to do my best in all matters of life. I especially want to thank my partner Yoly, for her infinite patience.

Table of Contents

List of Figures

List of Tables

Chapter One

1 Introduction

There are two main forms of interaction between humans living in urban areas and water as described by Butler & Davies (1999). The first is the abstraction of water from the water cycle to be used for living; this means for drinking water, industrial processes, etc, and the second interaction is related to the rainfall over the urbanized area. These interaction give rise to two types of water, of different quantity and quality, that have to be drained and conveyed through the urban area. These types of water are "waste-water" and "storm-water"; both types can be conveyed by the same system (combined system) or in a different system (separated system). Failing to provide a reliable way to drain excess water in either system could have several implications in terms of damage to property, human life, health, environment, daily activities, etc.

More than five billion euros per year are invested in Europe to avoid flooding or improve drainage networks in urban areas (Sægrov 2004). The population living in urbanized areas is growing rapidly around world as does the need for basic services, including drainage systems. With the increase in population, the need for funds to provide or maintain basic services is also increasing. Usually, there are insufficient funds to extend and rehabilitate drainage systems (Ole et. al. 2004) yet each year governments have to find funds to plan, build, maintain (rehabilitate) and operate their urban drainage systems. Organizations in charge of assigning funds to the sector, whether government and/or in the private sector, do not have infinite funds; they can only assign finite funds following plans suggested by technicians for a horizon of several years.

These plans must be elaborated by optimizing their use of the budgets. However, the achievement of an optimal use of available funds it is not a straightforward activity due to the existence of conflicting interests, such as minimizing rehabilitation costs, maximizing performance and improving reliability of the drainage system, minimizing flooding risk, etc. All of these are subject to constrained budgets and stakeholder preferences.

Another issue of importance is the distribution of funds which usually has a large degree of uncertainty. The process of searching for an optimal use of monetary resources is subject to many sources of uncertainty. From an economic point of view there are uncertainties not just in the value of money (inflation) but also in the physical processes that are involved in the drainage system, and the way in which they are modeled.

It is also well known that the climate is changing. This is one of the most important issues in the planning and distributions of funds for the long-term rehabilitation of urban drainage systems. What will the rain intensity be in two, five, ten or more years? Which pipe diameter should be installed today in order to minimize costs if the future is uncertain?

There are others source of uncertainty in urban drainage optimization; for example population and urbanized areas are increasing; this implies that there is a need to build new service networks and improve the old ones; but once again this is not a deterministic process. Managers must take decisions today with looking to the future; technicians and scientist must provide tools to help reduce or put bounds on such uncertainties.

1.1 Urban Drainage in Developing Countries

The definition of developing country is very wide; it is related with the degree of industrialization in the country. This is rather variable, being different from region to region, so the definition does not give much information about internal conditions of the country that can help to come out with general methodologies to solve specific problems; they are unique and procedure has to be flexible. Developing countries do not escape the urbanization problems; their chaotic growth makes it even more difficult to manage urbanization and does not allow storm-water management to be carried out in a proper manner.

If the population growth in urbanized areas in Europe and Latin- America, for instance, is compared; it can be seen that while Europe maintains a very small grow rate in urban areas, the Latin-American urban population grows at a higher rate   than Europe (Figure 1.1). Countries like Venezuela have a current urban population of 94% of its total population, and Brazil has 87% which is still growing. Usually people tend to settle close to main cities, where economic activities are large and give more opportunity for personal development.

Large cities, if not well planned, are very difficult to manage. Providing services such as electricity, water supply, sewer and storm-water drainage is difficult. If it is true that the growth of mega-cities, that is cities with more than 10,000,000 inhabitants, stopped; it is also a fact that 50% of such cities are located in developing countries. In Latin America some of these mega-cities are Mexico City (Mexico), São Paulo (Brazil), Buenos Aires (Argentina) and Rio de Janeiro (Brazil). Even in cities that are not mega-cities but are also large, management may be difficult. A city with more than 1,000,000 of inhabitant is large enough for service provision to be complicated. Figure 1.2 shows the percentage of people living in cities larger than one million for countries in Europe and South America. It can be seen that in South America a greater percentage of people prefer to live in a large city than in Europe. In this case Brazil and Colombia are at the top of the table. This urbanization occurs because people leave the rural areas which have low service cover to the cities, looking for a better life. Most of the countries in South America have changed their agricultural economy to one of exploiting their mineral resources. This kind of development leads to economic growth but low employment due to bad economical decisions, with agricultural goods imported at lower prices than in local markets thus promoting migration from rural areas to urbanizing areas.

In Latin America the cities used to have development plans, but drainage it is not considered necessary as an integral part of such a plan for the city. Most of the plans are old dating from 1980s. At that time cities were growing but at a lower rate, and they were more pleasant to live in. They used to have green areas and houses with land lots that help water infiltration, such that flooding was not a major issue. Most countries have not fully implemented their plans or followed them for a number of reasons.

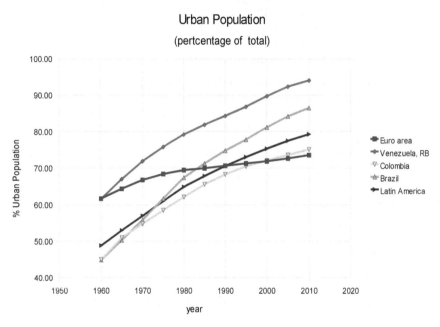

Figure 1.1: Urban population living in urban areas, percentage of the total population of the country (source:World Bank, 2011)

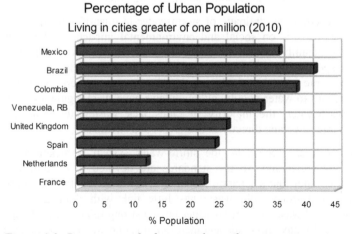

Figure 1.2: Percentage of urban population living in cities greater than one million (source: World Bank, 2011)

Among the main reasons for not implementing drainage plans is the fact that cities have grown in a very chaotic manner. People who come to cities from rural areas used to settle on the periphery of the city living in squatter settlements (called *"favelas"* in Brazil or *"Barrios"* in Caracas). For instance fifty percent of Caracas population live in this kind of settlements. These people build houses in areas that do not have any kind of services and drinking water and electricity become the main need for people living in such conditions, while sewers and drainage are not on the priority list.

Storm-water drainage in peripheral areas of developing cities usually do not exist; in the best cases only the main streams of the natural drainage remain. Governments do not require these people to follow the development plan for the city and prefer do not to apply the law. Common factors regarding storm-water drainage in developing cities are:

- Low investment in urban drainage facilities

- Increase of peaks flow due to the increase in urbanization

- Inadequate management of its land use

- Lack of a drainage system capacity and its maintenance in poor areas

- Poor design

- Overflows from combined sewer systems

- In separated sewer systems part of the sewage is directed to the storm-water drainage network

- Lack of control on flows as urbanization increase

- Human occupation of flood plains

- Lack of law enforcement in flooding areas, among others

Whereas in developed countries drainage of storm-water in quantity and quality is not a major issue, and the focus is on social and environmental impacts of the drainage, in developing countries quantity is still of major concern. Each year flooding in South America and, more specifically in countries like Colombia, Venezuela, and Brazil, leave people without their home and in particular causing damages to property, and even taking human lives. With regard to water quality, although laws have strict regulations concerning the treatment of wastewater, the majority of cities lack wastewater treatment plants (WWTP), and discharge sewerage directly to the rivers and streams.

In some countries such as Venezuela, the storm-water system is separated from the wastewater system by legislation, so the standards are devoted to dealing with water quantity. There is the belief that a separate system will avoid the contamination of storm-water. For instance in Europe solid waste has little link with drainage, however in developing countries solid waste and storm-water have a strong relationship. People living in poor settlements usually have no roads and therefore no solid waste collection service. It is common practice to ilegally dispose of solid waste in small drainage streams; this reduces their conveyance capacity causing yet, more severe flooding (Figure 1.3). Also, litter in the streets, and transit of old vehicles increase the negative impact of the water quality. There is little data in developing countries about the first flush effect and the pollutants that it may contain, which finally increases the pollution in receiving waters.

(a)   (b)

Figure 1.3: (a) Garbage accumulated in a drainage conduit in Belo Horizonte-Brazil (b) litter after a flooding event in Caracas-Venezuela

Any master plan for a city in a developing country has to address and develop the following points:

- Plan the urban flooding and storm-water drainage controls; it is more expensive to build such controls after urbanization than to make a plan before hand

- Institutional regulation; frequently there are not regulations for drainage, or they are out of date, and specifically storm-water drainage is not assigned to a specific institution or department

- Law enforcement; the problems of areas being urbanized by informal settlements has to be addressed; politicians avoid taking unpopular measures but it have to be addressed by the government administration

- Capacity building; professionals at all levels are required, urban planners and engineers need to be formed with the adequate concepts and provided with suitable tools to deal with the drainage problems

- Public participation; any plan has to be accepted by the final users and have their commitment. It is common that communities raise their concern about deficient drainage services during flooding events, but after some months it is forgotten by both the administration and users

While it is undeniable that developing countries lag behind the developed countries it is also true that there is no excuse for not facing new issues concerning drainage management. Landscaping, capacity building, climate change, flexibility in design, and stakeholder participation, have to be included, in addition to dealing with water quantity and quality in the management of drainage in developing countries. Additionally, most cities are relatively young but the drainage system start to be inadequate after 30 or 40 years when most downtown areas require the rehabilitation of their drainage networks.

1.2 Multiple criteria in Urban Drainage

Decision making when planning the development of urban drainage systems is in essence a multi-criteria task. It means that any approach oriented at dealing with urban drainage, design or rehabilitation, involves a variety of stakeholders in the selection of the best and most sustainable drainage design; the rehabilitation plan needs to consider not only water quantity but also water quality and amenity (Ellis et all 2004). The selected approach for developing the plan must be sustainable, and it must integrate in a balanced way technical, environmental, economical and social interests.

Sustainable urban drainage systems are evaluated taking into account the benefits that they may provide to the main three aspects that are considered in Figure 1.4. The first aspect is water quantity; this is oriented towards the mitigation of impacts due to flooding risk using traditional drainage, but the means to achieve the reduction in risk of flooding has to be in accordance with the aspects of quantity and amenity. The components for flood risk reduction tend to be mainly based on volume (storage) rather than on pipes (conveyance) as with traditional approaches.

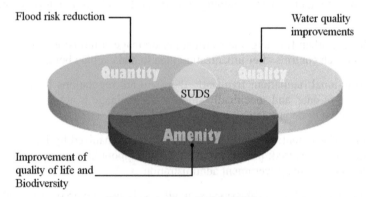

Figure 1.4: Sustainable UDS

While the quantity is important to reduce flood risk, water quality also has to be taken into account. The orientation in the use of extra volumes to hold abundant water and not only the pipes to drainage storm-water produces water of better quality and at the same time reduces flood damage. Capturing the rain water from the very beginning when it is fallen, limits water entering into contact with major sources of pollutants. Green rooves, water harvesting and infiltration in pervious areas limit contamination of the drainage water. Once the water reaches the urban areas and streets, it starts to change its quality. The use of wet ponds, soak-ways and WWTP can also help to improve quality. If good management is done an equal water quality can be expected in both, in water runoff and in the final discharge as well.

There are social values that complement or improve the quality of life of the inhabitants of a city. Cities with the land surface covered fully with concrete are not pleasant to people. Amenity is an important social value that is closely related to landscape and biodiversity. Wet ponds and wetlands can provide a nice landscape and sustain wild life; this provides a good aspect that can be reflected in an economic assessment. People are willing to pay more if they can live close to an attractive area with wild life.

It is undeniable that engaging in multiple targets makes the problem more complex because some of the objectives can be in conflict. An appropriate approach for urban drainage design or rehabilitation is to merge several fields in order to ensure a robust decision making process, providing managers and stakeholders with an effective decision support system that allows a satisfactory tradeoff between conflicting objectives (Lounis 2000). It is argued here that a multi-criteria approach is needed in order to include the participation of different actors and their interests by establishing a mean of safe negotiation for all the parties involved.

In order to have a sustainable storm-water design, there must be a synergy between environmental health, economic prosperity, and social values, andthese perspectives have to be seen through ongoing communication with stakeholders. Usually the social impact of a storm-water system is difficult to evaluate, but it can be very important. For instance the perception of users on the reliability of the system can influence the decision of buying a property and its sale price. In this case the social perspective overlaps with economic interests.

1.3 Rehabilitation of UDS

While the process of rehabilitation of urban drainage system will be addressed in a separate Chapter, a brief introduction is necessary in order to provide the reader with the context of the problem. First rehabilitation has to be defined to determine the scope of this thesis. Rehabilitation can be seen from two points of view depending on the drainage system evolution.

A drainage system degrades due to several factors. Aging is the principal factor. For instance the performance of pipes starts to degrade with time; they reduce their conveyance capacity due to an increase of the roughness of the pipe material, the rest of the assets also suffer from aging. Structural integrity of the system can also reduce performance. Pipes cracking under a heavy load can compromise the integrity of the system and reduce its effectiveness. In general the performance of a drainage system will be reduced in time.

If the demand on the system is constant an update of the system is required at an approximately constant interval of time (Figure 1.5). However the external demand is not a constant, it is dynamically changing over time. Factors such as urbanization and climate change are the most influential parameters that promote an increasing demand on the drainage system. In cases where both demand and capacity of the system are dynamic, it is necessary to upgrade the system. It not sufficient to return the drainage system to its original design capacity, it has to be rehabilitated on such a way it can deal with new demands; see Figure 1.6.

While developing countries may a simplistic approach to urban drainage design and rehabilitation, Europe has come a long way in these areas. Urbanization problems were faced since the 18th century, and in the 1960s committees started to develop manuals for sewerage network rehabilitation.

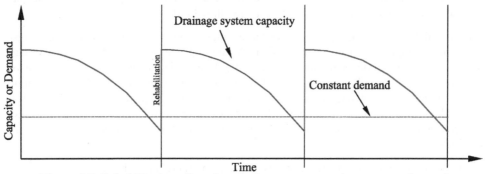

Figure 1.5: Rehabilitation of a urban drainage system under constant demand

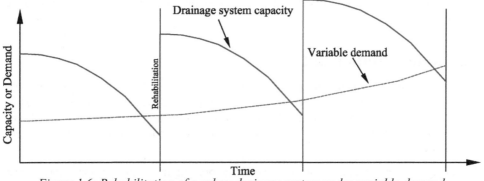

Figure 1.6: Rehabilitation of a urban drainage system under variable demand

The rehabilitation process is not straightforward; each country, and more specifically each city, has their own internal factors that influence the final procedure to be followed. It is common that each city or municipality in Europe and America has its own urban drainage manual; however, they also have to follow some directives coming from the central administration that contain policies and general aspects of the urbanization process. Nevertheless there are several common features or steps that must be followed in one form or another. Three main stages can be identified in an rehabilitation plan for urban drainage systems. They are:

• Diagnosis

• Assessment of environmental, structural and hydraulic conditions

- Development of the plan

First, diagnosis must be done. This step consists in the identification of critical pipes and areas, and it can be done using past data records. Before carrying out an effective diagnosis, it is necessary to choose criteria and indicators which are aimed at measuring the drainage system performance.

An assessment of the current system performance can be done with the measured data, but assessment of the future performance based on the improvements in the network has to be done with computer based mathematical models. Models are extremely important for the *"what-if"* (scenario) analysis necessary for any evaluation of environmental, hydraulic and structural conditions of UDS.

Finally, the rehabilitation plan has to be developed by developing and evaluating different scenarios using a multi-criteria approach. There is no easy way towards the unique solution for the rehabilitation process. Indeed, several solutions can satisfy different stakeholders with different interests that are usually conflicting. A set of solutions is the best way to provide a wide spread of alternatives that can match stakeholder needs and open up negotiations between them, with the final objective to develop a consistent plan. In this sense drainage system rehabilitation has two key aspects to consider: one is the development of appropriated hydrological and hydraulic models, and the other is the formulation of indicators of performance. Both are addressed in the following sub-sections.

1.4 Performance Indicators

In order to proceed with the development plan and the multi-criteria approach, a set of good performance indicators have to be used. Selection of appropriated levels of service is a critical task if optimization process is planned for the generation of alternatives and their evaluation; they contribute to defining the objective functions to be optimized. A wrong selection of the performance indicators can produce redundancies in the objectives, and generate additional computation and time costs. They can also lead to generation of unclear or meaningless decisions.

The most common indicator or *"level of service"* used is based on the notion of the frequency of flooding. A level of protection is established defining the probability that a particular event is exceeded (return period). Also this probability combined with cost or damage function can be viewed as a risk. However, the concept of level of service should be wider: the damage done by flooding a park is less than that of flooding a hospital. The influence of land use should therefore be taken into account.

In order to protect urbanized areas from flooding, not only structural measures can be used, but also non-structural measures can be employed to solve flooding problems. For instance, a change in land use can be one of these non-structural measures. This means that land use can be a variable to be included in the evaluation of scenarios. Another aspect to address is the type of infrastructure. For example, some buildings are more resistant to flooding damage due to water depth and/or flow velocity, depending on the construction material. The *"real"* concept of risk is not just the probability of flooding times some kind of cost; the vulnerability of structures has also to be taken into account in such an analysis.

Additionally, levels of service to describe social and environmental issues have to be properly defined and combined, in order to provide decision makers with suitable tools to determine an appropriate expected cost of each measure or scenario, and to address appropriated budgets for the rehabilitation. Also, these indicators will complement the multi-criteria analysis.

1.5 Role of Hydraulic Models and Optimization Algorithms

It can be argued from the stages cited above that hydraulic models play a key role in the rehabilitation process. The interaction between below-ground (pipes) and above ground (streets, canals, rivers) has not been well explored. The problem of surcharge in pipes is not still properly implemented in existing modeling software. The conjunctive use of 1D and 2D models is also a matter of current research (Mark 2004). In order to carry out an evaluation of performance and the estimation of indicators, it is necessary to use hydrodynamic models.

MIKE URBAN by the Danish Hydraulic Institute (DHI) includes 1D modeling components, and allows for coupling with the 2D models (e.g., MIKE21), which can jointly model overland and pipe network flow. Sobek 1D-2D by Deltares, is also a suitable tool for above and below-ground simulation. However such coupling is not straightforward and can exhibit instabilities. There are also other models for 1D or 2D modeling but few deal with integrated above and below-ground modeling.

Hydraulic models play a major role in the optimization process; in particular they transform the input data (rain and network topology) to generate flood levels, velocities and discharges. These results are post-processed as inputs to cost functions, which are evaluated in the optimization algorithm. The computational cost and the number of evaluations are considerably important. These will determine the applicability of the optimization process in practical cases.

The evaluation of possible scenarios implies the use of optimization techniques. Due to the multicriteria nature of the sustainable urban drainage approach, multi-objective optimization (MOO) can be used as an effective tool to merge different objectives and stakeholder interests or needs. Several tools based in heuristic method as Evolutionary Algorithms (EA), and Random search exist; however, they are inevitably computationally intensive since require a considerable number of (complex) model runs and this may be a bottleneck.

This prompts for developing new, more efficient algorithms of multi objective optimization. They need to exploit the particular characteristics of the storm network system, reducing the time needed for the MOO process. For single objective optimization, a number of algorithms has been developed that show better performance in time than canonical EAs (e.g., ACCO by Solomatine 1998). (In this work we test an approach to MOO based on a sequence of single objective optimization runs that outperforms a widely used MOO evolutionary algorithms in terms of the needed computer time, which is very important when one deals with complex UDSs.)

1.6 Objectives of the Study

As can be deduced from the brief presentation above, the rehabilitation of an urban drainage system is a very complex task that involves several fields of knowledge and a range of stakeholders, whose interests can be conflicting depending on the roles that they play in society. Decision makers need suitable tools that allow them to simplify the decision making process. There are some approaches for urban drainage rehabilitation like CARE-S (2002-2005), or SWITH (2006-2011) that address the problems of urbanization in cities. However, some of these methodologies have not yet been fully tested or are rather complex what makes them difficult to apply.

It is argued that a decision support system framework that takes into account the use of multiple objectives and uncertainties, constrained by available budgets and involving urban planners, engineers and other stakeholders, is needed to facilitate urban drainage decision-making to allocate funds in an optimal way. Such a framework must be reliable and fast in terms of processing speed. As mentioned above the evaluation of the objective functions (service performance indicators) requires the running of hydrodynamic models, which are computationally intensive; looping them inside optimization algorithms will require yet more computational power.

The present research is aimed at establishing a framework as a base for a future DSS tool in order to optimize the allocation of funds for the rehabilitation of drainage systems through the evaluation of different scenarios; and at investigating to what extent it is possible to reduce the computational time in order to improve the optimization algorithms for application in real problems. One issue which makes the optimization process in real world application difficult to apply is to come up with a feasible solution in a suitable time for practitioners. The reduction in computational time will therefore make the results of the research useful to practitioners. A selection of several levels of service for evaluation will be tested and compared. Another desirable characteristic of the framework to be developed is that it has to be suitable for application at developing countries.

Based on the above, the following *main objective* of this research can be formulated: to develop and test a framework for the rehabilitation of an urban drainage system in the context of a multi-criteria approach. The *specific objectives* are:

- To review the state of the art in rehabilitating UDS

- To review the state of the art in multi-criteria optimization

- To develop and test a model-based framework for the rehabilitation of an urban drainage system based on the use of multi-objective optimization algorithms

- To explore the possibilities of reducing overall optimization time using different algorithms and methods of optimization

- To develop a parallelized multi-objective algorithm based on the multi-core and cluster computation with the aim of reducing the computational time and/or increasing the size of the optimized urban drainage networks

- To test the framework on real study cases for the cities in developing countries.

1.7 Outline of the thesis

The thesis is organized in eight chapters. Chapter One provides introduction, motivation and objectives of the study.

In Chapter Two the state of the art in urban drainage rehabilitation is reviewed. Current methodologies suggested in the European Water Framework Directive are also reviewed. Emphasis is given to experience in the UK which has one of the most extensive set of expertise in sewer rehabilitation. Several indicators of performance are reviewed and classified. This chapter also reviews hydrological and hydrodynamic models applied to in drainage context.

Chapter Three is devoted to the new drainage rehabilitation framework. This explains how a multi-criteria approach is taken into account and how expert knowledge is incorporated in the process of scenario generation for evaluation. It is also described how the integration components and hydrodynamic tools is coded into a prototype called OPTRESS.

In Chapter Four optimization tools and algorithms are explored, starting with traditional optimization methods such as linear and dynamic programming, and single global optimization using population based algorithms. Non-dominated Sorting Genetic Algorithm (NSGA-II), and epsilon Non-dominated Sorting Genetic Algorithm (ε-MOEA) are reviewed and recoded under Windows OS (NSGAX system). Also a new approach using a clustering algorithm is proposed and tested to be employed in Multi-objective Optimization using Single-objective Search (MOSS) for computationally demanding problems.

The testing of the proposed framework is done in Chapter Five. Performance indicators and objective functions are defined. A comparison of the multi-objective algorithms is done with reference to an urban drainage network. Stopping criteria for the optimization are evaluated. Sustainability takes into account the use of wet ponds. Also intangible costs are considered in an example with three objectives.

Chapter Six deals with the use of parallel computing. It explains how the NSGAX is parallelized in order to deal with computationally intensive optimization problems and to reduce time taken for urban drainage rehabilitation projects. A real application is done for a sub-catchment of Belo-Horizonte city in Brazil.

Chapter Seven describes two more real applications of the framework, both in developing countries. The first is a city in Venezuela. In this case the aim is to rehabilitate the drainage system of the Cabudare city taking into account economic values. An expected value of damage is calculated and compared with a single return period solution. The second study case is in a sub-catchment in the city of Cali in Colombia. This application includes water quality and optimization is performed for the three objective functions.

Finally, the conclusions and recommendations are presented in Chapter 8.

Chapter Two

2 Urban Drainage Rehabilitation

Urban drainage rehabilitation is a complex matter; it involves a combination of several subjects in order to develop a sustainable plan. Hydraulic, economic, environment and social assessment is required to achieve this goal. This chapter is focused to review the state of the art on the drainage rehabilitation subject. Critical pipe identification, rehabilitation techniques, hydrologic and hydraulic modeling, performance indicators are part of the items to be reviewed. Also, the existing methods and standards for rehabilitation of drainage systems are reviewed.

2.1 Introduction

The fact that the urban drainage rehabilitation process is not a straightforward task can be explained by the dynamic and integrated nature of an urban catchment and the complexities associated with multi-objective decision making. In the past, the operation and maintenance of drainage systems were carried out in a manner that did not give much consideration to the optimal expenditure. Many rehabilitation projects lack a cohesive plan and are based on the strategy of "clean as many inlets as you can afford and fix the system when it fails" (Fenner 2000). However, since the mid 1970's engineers and planners started to realize that system renewal and maintenance costs were increasing and that industry practices had to be improved. With the introduction of modern asset planning practices in the 1980's, new approaches based on the optimal system management concept started to emerge. In the 90's the concept of sustainability was introduced with the use of a series of techniques called best management practice (BMP). Asset management (AM) may be defined as a comprehensive and structured approach to the long-term management of assets to serve the needs of urban communities at the lowest possible cost (Roland Price and Vojinovic 2011). The AM approach encompasses management, financial, customer, engineering and other business processes. The reason why AM is considered a business discipline is that its success is measured using business and financial indicators. It is also important to note that asset management encompasses a large number of diverse processes including engineering, but many are not engineering in nature.

In order to carry out a reliable drainage rehabilitation plan it is necessary to bring together several fields and tools. Knowledge about why the rehabilitation is needed, the type and causes of failures, identification of critical areas and critical sewerage assets, are essential in the rehabilitation process. Current rehabilitation techniques, the sustainability of the drainage system, modeling tools, and optimization techniques among others, have to be included in the rehabilitation in order to achieve the best solution.

It is important to define what we understand by rehabilitation. Drainage network rehabilitation may be needed for several reasons. Structural failure of the network components is the major factor. Pipes, channels tanks, and other sewerage facilities are affected by aging, over loading, chemical damage and other factors that reduce their nominal life. But they are also affected by continuous changes to the city drainage network. Factors like growing population and urbanization, changes in land use and climate can introduce changes in the serviceability of the drainage system, even if it retains its initial conditions. A frequent question is: does rehabilitation consist in returning the system to its initial condition or does it include an upgrading process?.

In our case we include in the concept of rehabilitation the need to upgrade the system. The failure of the system to deliver or perform as it was designed and as it was intended implies the need for rehabilitation; therefore, the failure of the performance and its effects on public health and the environment are signs than a rehabilitation procedure has to be carried out.

For strategic sewerage asset management including rehabilitation the following issues are of paramount importance and have to be addressed:

- What is the current condition and performance of the assets.

- What is the required level of service of each asset to meet the requirements from the public, regulations etc..

- What is the current performance of the overall system.

- What are the critical assets to sustain that performance.

- What are the best rehabilitation options in terms of investment in order to maintain the performance of the system.

Finally, mathematical models play an important role in the rehabilitation of sewerage networks. It is almost impossible to carry out a rehabilitation procedure without the use of hydrological, hydraulic, and quality models.

2.2 Urban Drainage Asset Management Cycle

Several approaches that aim to reduce rehabilitation costs have been developed. The most well known is the approach developed by UK in the 1980's; see Error: Reference source not found. The complete rehabilitation process has been divided into several tasks or phases. The results of each phase are analyzed, and depending on such analysis, the practitioner decides whether to continue onto the next step or stop until an established criterion is met in order to continue to the next phase.

Initial planning starts by defining the standards of performance for the sewerage system. Assessment of the performance of the existing system is therefore an essential first stage, and a fundamental part of rehabilitation planning (http://www.hi.ihe.nl/srguide/). These standards relate to public health, flooding risk, structural integrity, economics, etc. The standards require indicators to reveal how good the system performance is. They are designed to act as quality service benchmarks. In this phase an initial assessment of the performance is made, using past records of flooding, sewer overflows, pipe collapse data and other related information.

In Phase one (1) a preliminary diagnosis of the sewerage network is carried out. Critical parts of the network are identified using the existing records. Critical areas are defined depending on the cost and extent of the damage that would occur if the system collapses at any point. If there are not enough records to identify such areas, the records must be upgraded in order to meet the minimum requirements.

Phase two (2) is aimed at assessing structural, environmental and hydraulic states of the network. Assessing the structural condition leads to the elaboration and application of a inspection plan, from which parts of the networks at risk are identified and the total length of pipes to be rehabilitated is determined. The impact on the environment is carried out also in Phase two (2). The risk of pollution and damage to the environment are assessed. Water quality planners are in charge and work in collaboration with hydraulic engineers. Assessing hydraulic performance involves investigations of the whole network, building and verifying the hydraulic model, using the models to perform simulations evaluating the performance of the network for different scenarios, and identifying the location and cause of performance deficiencies.

Phase three (3) deals with the comparison of the scenarios. A set of potential solutions to the problems identified in Phase two (2) (structural, environmental and hydraulic) should be posed and evaluated. A priority should be assigned to the different solutions and the less expensive scenario should be identified. The availability of funds should be also taken into account .

Finally, the plan should be implemented. This consists of the design and construction of rehabilitation works. Continuous monitoring is then needed and the hydraulic model should be kept up to date. When new information is acquired and/or circumstances change a review of the plans has to be done.

These steps are very general. New technologies for rehabilitation, modeling, and evaluation are constantly under development and this general approach allows their incorporation in the rehabilitation plan.

2.2.1 Levels of Service and Indicators

The definition of levels of service or performance indicators is a critical issue. These are used to assess the performance of the drainage system. The concept of a level of service for a drainage systems is usually restricted to risk; for instance, the number of times in which a property is flooded once or two in a period of ten years (OFTWAT 2006). However, its definition can be interpreted from a wider point of view: like any grade of satisfaction or agreement about the service provided by the system. It can be measured in quantitative or qualitative form .

There are two kinds of performance criteria, namely fixed and variable. Fixed regional values come from experience. Such an approach is used when a variation in performance implies smaller increments of costs or when there is a lack of information. Variable criteria of performance involve the search for a level of performance that optimizes the balance between benefits and costs. This variable approach is applied when changes in level of

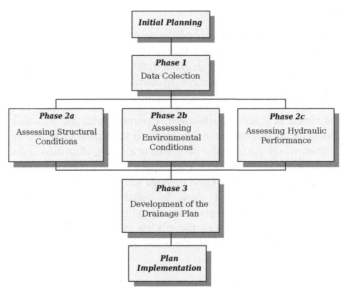

Figure 2.1: Urban drainage rehabilitation cycle (UK Sewerage rehabilitation manual)

performance increase costs and also when there exist other considerations such as social or political.

Indicators

The rehabilitation of the UDS must be done incorporating sustainability. This means taking into account the different actors and stakeholders, including their conflicting interests. Sustainability is based in the interrelation between technical, environmental, economic and social issues. Sustainable criteria as presented by Ellis et al. (2004) are composed of primary, secondary criteria and benchmark standards.

Primary criteria (Table 1) are the fundamental parameters for multicriteria decision-making. These primary criteria are founded in the four pillars of sustainability: technical, environmental, social and economical categories.

These primary criteria are very general; for that reason primary criteria are divided into more specific descriptors called secondary criteria. However these secondary criteria are only system descriptors without any measurable objectives. In order to really quantify the criteria a set of technical indicators is needed; Ellis calls them benchmarks.

Benchmarks must be measurable and could be standards or indicators. For instance, on the category *"Technical Performance"* a primary criteria is the system performance quantity. A secondary criteria for such enter, could be flooding performance. However, flooding performance can be assessed by several benchmarking indicators such as the number of floods per annum, number of properties affected, reparation costs per annum, etc. All these indicators, to a minor or major degree, are capable of measuring the performance of the system.

The selection of the indicators is done by decision-makers taking into account the stakeholders preferences. If the urbanized area is heterogeneous regarding the number of properties per hectare, the use of the number of floods per annum is not a good indicator of performance. Frequently, flooding over sparsely populated areas causes less damage that in crowded areas. In such cases it is better to use the number of properties affected per annum as a benchmark. Tables from 2 to 4 show some secondary indicators and their benchmark values. But they are just an examples, and more can be added depending on the practitioners', decision-makers' and stakeholders' experience.

Table 1: Primary criteria for assessing SUDS sustainability (Ellis et al. 2004)

Sustainability criteria Category	Primary Criteria
Technical and Scientific Performance	System performance (quantity and quality) System reliability System durability System flexibility and adaptability
Environmental Impacts	Water volume impact Water quality impact Ecological impact Resource use Maintenance, service provision and responsibilities
Social and Urban Community Benefits	Amenity, aesthetic, access and community benefits Public information, education and awareness Stakeholders acceptability, perception and attitude to risk and benefits Health and safety risks
Economic Costings	Financial risk Affordability Life-cycle costs

Table 2: Benchmarks standards for the Technical & Scientific performance

Primary Criteria	Secondary Criteria	Benchmark/Standard Indicator
System Performance (quantity & quality)	Storage & flooding	Design storm, Storage volumes number of flood/annum, number of properties affected Downstream protections, Disruption costs/time, Pollutant concentration, probability of exceedance
	Receiving water	First flush capture potential Complains per year Events captured for treatment Pollutant degradation rates
System Reliability	Performance reliability failures: health and safety	% Pollutant removal, Eutrophic status, odorous, stagnant water and bacteriology Risk of failure, Operational safety
System Durability	Design life	Operational life time, storage volumes, sediment accumulation rates
System Flexibility	Capability for change and retrofitting	Design free-board (storage and quality) Easy of retrofitting and/or add-on structures and features

Table 3: Benchmarks standards for the Economic Costings classification

Primary Criteria	Secondary Criteria	Benchmark/Standard Indicator
Life-cycle costs	Investment and Operational Costs	Design and capital costs Operational and maintenance costs Plant replacement costs Sediment monitoring and disposal costs Site decommissioning costs
Financial risk	Risk exposure	Cost befits analysis Investment loss risk Operational health and safety risk Site reclaim value
Affordability	Long-term Affordability	Adoption and liability coverage Economic add-on value (enhanced land/property values) Amenity income streams Long term-management provision and costs

Table 4: Benchmarks standards for the Environmental and Social classification

Primary Criteria	Secondary Criteria	Benchmark/Standard Indicator
Water volume, quality and ecological impact	Pollution control	Groundwater quality contamination Oxygen concentration, Number of overflows, Overflow volume, Overflow duration
	Ecological diversity	Biological indexes, Willingness to pay, Aquatic live in the stream
Amenity, aesthetic, access and community benefits	Social inclusion and multi-functional use	Amenity level, Education level, Preparedness level, Normal life disruption
Health and safety risks	Home contamination Infections and mental diseases	Cases of asthma, diarrhea, coughs, kidney infections, panic attacks, viral infections, stress disorders etc.

2.2.2 Criticality

Engineers realize that only a few percentage of failures consume a high proportions of their budget. Peters, cited by (Fenner 2000), reported that 20% of the failures produce a disproportionate amount of the repair costs in the UK. This fact reveals the need for the use of selective strategies in rehabilitating urban drainage systems. There are mainly two (extreme) approaches for the rehabilitation of urban drainage: one is to do "*reactive maintenance*" and the other is to do so called "*preventive maintenance*". It can be argued that the asset failures which produce the bigger financial, economical, and social costs (20%) needs a preventive maintenance plan, while the rest (80%) can be approached using a reactive plan (Butler and John Davies 2004).

The criteria used to prioritize inspection and rehabilitation should include not only the physical conditions of the pipes, but also the degree of impact of a sewerage failure (McDonald and Zhao 2001). The impact of sewerage failure can be determined taking into account various aspects:

- *Asset Location*: This is related to how the public and environment are affected if a failure occurs and it depends on several regional aspects. Land use around the damaged asset determines how many person are affected and the monetary cost of the damages. High traffic load in the proximity of a flooded area leads to an intangible cost like delays, people stress, and other disturbances. Proximity to critical establishments such as hospitals, educational buildings, government, airport or military structures contribute in the criticality ranking. Failures of assets located downstream are more critical than assets located upstream because the drainage area is larger.

- *Soil support*: The interaction between the soil and the buried sewerage assets plays an important role at the moment of deciding the criticality of an asset. A combination of high hydraulic heads in pipes can cause ex-filtration promoting the development of voids in the soil, leading to fractures in the pipes and surrounding structures. Pipes laid on sand have a better rank than pipes laid on a high plasticity material like clay.

- *Pipe size*: This gives an indication of the relative importance of the pipe. It is expected that large sizes of pipes are located downstream on the system and a failure of such pipes induces a failure of the whole system. Also they give an indication of the magnitude of the rehabilitation work.

- *Pipe depths*: The deeper pipes are difficult to maintain and rehabilitate. Failures will need more effort and time to repair, leading to major and larger disruptions in the area increasing the severity of the damage.

- *Areas of seismic activity*: Seismic movement can produce a fracture in sewerage assets. Cities used to be divided into regions depending on the acceleration in the soil during a seismic event. The type of soil and magnitude of the movements are very important issues. Together they can produce an uplift on sewer pipes caused by the liquefaction of the surrounding soil.

Another important factor to be introduced is the implementation sequence. When an urban drainage network is rehabilitated the spatial and temporal order of the rehabilitation is very important. If the rehabilitation starts from upstream a larger discharge is expected downstream because the hydraulic conductivity upstream is improved. If the system downstream has not the capability to convey such a discharge, the system downstream will fail leading to a larger flooding damage. It is better therefore to start the rehabilitation from downstream to upstream.

As is stated by Butler and Davies (2004), a critical sewer usually has the following characteristics:

- above-average depth (3 meters or more to invert)

- brick or masonry construction

- large diameter (above 600 mm)

- bad ground (surrounding soil)

- under a busy traffic route

- under buildings, railways, tram routes, canals, rivers, main shopping streets, primary access to industrial zones, motorways, etc.

- with difficult access if failure occurs

2.3 Sewerage Rehabilitation Assessment

Once the system is inspected, available data is collected and recorded, indicators and critical areas are defined; the next step is to check the conditions of the assets. The approach should focus on the following three main aspects:

- structural integrity or physical conditions

- hydraulic adequacy or capacity

- environmental and social conditions

Establishing the conditions of sewer pipes requires a good judgment or experience in addition to an objective technical analysis. Since it is virtually impossible to predict the precise moment a pipe will fail or collapse, categorization using different levels of deterioration (or risk) make it possible to rate the condition of the system.

There are a number of sewer condition ratings that have been developed as general guide lines. The most well known have been developed by the Water Research Center (WRc) in Europe, the Water Environment Federation and American Society of Civil Engineering (USA), and the National Research Council, Canada. Although most of the municipalities in developing their rehabilitation plans have adopted a specific condition assessment system, others have adapted these guidelines. The following documents, manuals and guidelines have been widely used:

- Manual of Sewer Conditions Classification (WRc, 2004);

- Guidelines for Conditions Assessment and Rehabilitation of Large Sewers (IRC,2001)

- Assessment and Evaluation of Storm and Waste-water Collection System (NRC,2004)

- Manuel de standardisation des observations-inspections télévisées de conduites d'égout (CERIU 1997);

- Existing Sewer Evaluation & Rehabilitation (ASCE, 1997); and

- Manhole Inspection and Rehabilitation

It is not the intention of this thesis to reproduce the complete contents of these manuals. It makes an overview regarding what could be included in the development of a DSS for urban rehabilitation. Not all the items can be implemented for all the cases; each city has its own requirements and particularities. However, the main philosophy remains applicable.

2.3.1 Assessing Structural Conditions

In order to carry out the structural analysis for the buried drainage system, it is necessary to understand the mechanisms that produce pipe failures. They usually start with a small defect, like a small crack due to an excessive vertical load or poor pipe bedding. Once the crack is formed in the pipe (or in the pipe joints) a leakage can promote soil erosion decreasing soil resistance, which can increase the crack size or promote the failure of other sewerage assets.

Defect size, infiltration and soil loss

In general terms, it is thought that the typical process involves the movement of water from the exterior of the pipe to its interior, or vice-verse, through defects in the sewer wall, and the transport of soil into the pipe. The main factors that affect the rate of soil loos are thought to be (Davies et al. 2001):

- sewer defect size;

- hydraulic conditions: water table, frequency and magnitude of storm;

- soil properties, among others.

Larger cracks produce more infiltration which promotes changes in the soil properties. Rogers (1986) carried out experiments referred to as "*small*" and "*large*" scale tests. The small-scale tests used an apparatus which consisted of a rectangular box of 600x300x150mm with an adjustable slot in the base to mimic a sewer pipe fracture. Soil was packed into the box and water pressure applied to upper surface of the soil. The large-scale test was built to simulate a pipe cross section. It consisted of a rectangular box made of two transparent sides 1440x2226 mm connected by a base and sides 200mm deep. A 150mm diameter pipeline ran through the box parallel to its short side. The pipe had a longitudinal crack in the crown and was surrounding with soil. This last apparatus allowed the water level to vary in the soil (water table) and the level inside of the pipe (surcharge) was given the possibility to have flow in both directions through the crack.

The results of these experiments showed that pipes with severe defects (> 10 mm) or larger defects (5-10mm) have a high risk of ground loss, while medium (2-5 mm) or small defects (<2mm) have a low risk of soil loss. Frequent surcharge in pipes with a high or low magnitude of surcharge have a higher risk of soil loss compared with pipes that have occasional surcharge or are never surcharged. The influence of soil type generates a high risk of soil loss for soils with plasticity index less than 15 and a low risk for a plasticity index greater than 15, ranging from high to low in following order: silts, silty fine sands and fine sands; low plasticity clays; medium to coarse sands; fine to medium gravels; well graded sandy gravels; medium to high plasticity clays; all clays if the sewer is constructed by tunneling.

Brick sewers are the oldest and more vulnerable sewers in network systems. They suffer loss of bricks due to the general loss of cement. Deterioration starts with a loss of mortar promoted by other problems like root intrusion, infiltration and sedimentation due to fine soil penetration. Dynamic forces can produce damage or deformation in underground pipes. Heavy surface construction, earthquakes or land slides, expansive soils can produce cracks in rigid pipes or their joints. Structural failure mechanisms can be found in Serpente (1994) (Figures 2.2, 2.3 and 2.4), Marsalek et al. (1998), Water Research Centre (2004), among others.

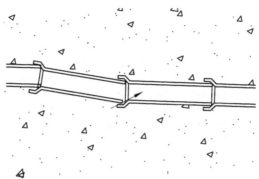

*Figure 2.3: Displaced joints in a concrete pipe
(Serpente 1994)*

*Figure 2.4: Mortar loss and infiltration on
bricks conduits (Serpente 1994)*

Corrosion

In tropical and warm zones the presence of hydrogen sulfide in sewage reacts with condensed water on the pipe crown forming sulfuric acid which corrodes the pipe, produces bad odors, damages sewerage assets and is a risk for personnel in charge of sewerage maintenance.

The process starts with the settlement of sediment in the sewer. Sediment is any type of particle capable of forming a bed deposit in sewers and other related hydraulic structures. Some of these particles are very small and are transported through the sewer as a *"wash-load"*. They have no major impact on the sewer transport capacity. However, they are very important for the pollutant loads carried to treatment plants or through overflows into urban stream.

In the other hand, larger (or dense) organic or inorganic particles which have settling velocities in the range between 0.2 m/s – 0.3 m/s can only be transported by larger rainfall events which have low frequency. Some times, they may form permanent deposits in the sewer network. These deposits are is able to retain finer sediments forming a bacterial slime in the sewer.

Usually waste-water is free of hydrogen sulfide but contains sulfur as inorganic sulphate or organic sulfur compound. Also, the sulphate comes from industrial processes, excreta, and ground water infiltration. The sulfur and sulphates are transformed rapidly into hydrogen sulfide as the product of bacterial activity contained in the sediment bed. The decrease in dissolved oxygen, a slow water movement, high temperature, and long detention time promote dissolved sulfide formation.

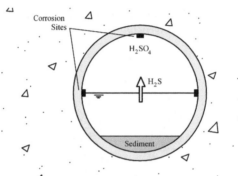

Figure 2.5: Corrosion in sewer, vulnerable places (Buttler and Davis 2004)

Turbulence in the sewer flow and high concentration of hydrogen sulfide causes the corresponding gas to be released from the waste-water, which produce bad odors. Another type of bacteria utilizes the hydrogen sulfide and water vapor to produces sulfuric acid. The sulfuric acid attacks the cement. Sewer flows can then remove softened cement, exposing aggregate particles to removal by the flow. This successively exposes new surfaces to attack, and eventually progresses to the steel reinforcement that is also attacked by the acidic environment (Chapman and Frank 2005). Vulnerable parts of the pipes are shown in Figure 2.5.

Asset aging

Aging is one of the main variables to be taken into account. It is expected that the risk of failure of a pipe increases when its age increases. As it was stated in Section 2.2.2, 20% of pipes are called critical. These critical pipes are the subject of a proactive maintenance program due to the high costs that a failure can produce. The remaining 80% of the pipes are under a reactive maintenance plan; their aging has more impact due to the lack of maintenance.

The classic survival function related with aging is called the "bathtub" curve as shown in

Figure 2.6 (Najafi and Gokhale 2005). At the beginning, when the pipe is just starting its life cycle, there is a high risk of failure. This is mostly due to human factors, like a lack of quality control in the construction of the bedding and joints, and manufacturing faults. These failures appear during the early part of the life cycle. Then a stabilization period is evident, and the risk of failures increase gradually until the end of the life cycle is reached; at this time an exponentially increase of the number of failures is expected.

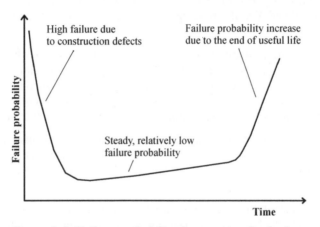

Figure 2.6: Failure probability due to aging (bathtub curve)

Structural Diagnosis

Structural condition assessment usually proceeds from visual inspection. It can be done by direct human visual inspection. For instance, walking along the sewer(if it is big enough) could help to identify the problems (Delleur et al ,1998). Sunken areas, pounding of water, deteriorating conditions at stream crossings, missing or broken manhole frames or covers, and deteriorating or failure conditions of catch basin inlets and other visible structures are signs of structural sewer deterioration. Soil maps, traffic load data, water burst incidents, borehole logs, property age data, sewer collapse records, hydraulic models and street works records are all useful to help in the structural diagnosis.

Large diameter sewers, greater than one meter, can be inspected using visual human inspection. However, the most common method to acquire the necessary information is the use of TV cameras (CCTV). CCTV images show pipe deformation, cracks, root intrusion and infiltration; they provide a diagnostic of the internal surface conditions. There are two main forms of CCTV installations, one is the use of a stationary system and the other is the use of a robot mounted camera.

Stationary CCTV can be installed inside critical or strategic pipes in order to inspect sewer lines and water-mains on-line. Also, they can be temporarily mounted at manholes and use their zooming capabilities to search for defects inside the sewer system. Defects close to the manhole can easily be identified, but the farther away the defect, the harder it is to identify and evaluate.

The use of stationary CCTV is cheaper and faster to carry out than the use of robots. It only requires the mounting of the camera in the manhole and does not require previous cleaning or emptying of the pipe for the inspection. However, the debris layer on the invert can hide defects, cracks or infiltrations. It is suggested therefore that the use of this technique in combination with other techniques that can give more detailed information of the structural condition of the sewer (Makar, 1999). First is advisable to carry out a scan with stationary CCTV and identify those parts of the sewer system which require a deeper study.

CCTV examination using a mobile camera system is the better approach. The camera is placed inside the pipeline and is either winched or self-propelled through it. Before running the camera inside the pipe, the pipe must be cleaned and its walls washed in order to expose the defects. The CCTV system can provide valuable information on the wall surface integrity; however it does not give information about the condition outside the sewer asset.

In order to evaluate the state of the material surrounding the sewer assets other techniques are required. The most reliable but costly is to dig and make a visual inspection of the surrounding soil; however non-destructive techniques like geophysical techniques are more appropriate in most cases.

Ground penetrating radar (GPR) is a non-destructive technique used to discover defects in the ground surrounding the buried pipes. The equipment transmits short radio wave pulses. Such waves travel through the soil, and when they find a soil change or any buried object they are reflected back to the transmitter. The interpretation of the returned waves gives a description of the soil properties. Materials that are highly conductive (with high dielectric constants) like metals rapidly attenuate the wave pulses and give zero penetration while in air the waves can travel very long distances. Sand, asphalt, and clay fall between these ranges.

GPR can be used from the ground surface or from the inside of the pipe and is capable of detecting voids, rocks, or buried objects in bedding. Also it can be used to detect ex-filtration or delamination in pipes walls. Disadvantages are that the method is highly dependent on soil conditions, results interpretation is difficult and requires an highly qualified expert, it can produce false positive results, it is more expensive than CCTV, a field test and calibration is required. Some authors have reported poor efficiency for pipes buried in clays (Hunaidi and Giamou, 1998).

Another alternative is the use of sonar or ultrasound systems. The inspection is performed using very high frequency sound energy. The sound travels through the soil or media, reflecting any change in the density of the travel media. Sonar can be used to work above or below the water surface; for that reason it is not necessary to empty the pipe system for the test. However, the transducers are not the same if the media is air or water. Sonar can be used to analyze the wall of the pipes, allowing the identification of cracks as fine as 5 mm. Also sonar has been used in the measurement of plastic pipes deformation (Wirahadikusumah, 1998). Some disadvantages are that it is somewhat more expensive than a conventional CCTV, a thorough cleaning of the sewer is require for better results, and parallel cracks are difficult to identify. It has been used in combination with CCTV images as a complementary source of information (Elmer, 2009).

Infrared thermography is another option than has been tested in order to assess sewer conditions. Thermography is based on heat transfer. Energy is transferred from warm areas to cooler areas allowing the acquisition of information about the surfaces of the objects. Depending on whether an artificial heat source is used or natural sources are sufficient, two different procedures, active or passive, can be applied. Subsurface defects such as deteriorated pipeline insulation, leaks, or voids can be discovered. The main disadvantages is the difficulty of interpretation, and the high volume of information. It is advisable to combine it with other kind of sensors such a CCTV or GPR (Duran, 2002).

Mechanical techniques can also be applied. These techniques are oriented to find the residual structural resistance of the pipe or to determine the condition of the mechanical element. They involve the use of hydraulic or manual tools to create a pressure on the pipe walls and to measure the reaction (deformation) of the pipe or surrounding media or bedding. Among mechanical techniques are: micro-deflections, the natural frequency of vibration, the impact echo and the spectral analysis of surface waves (Makar, 1999). The major limitation is to determine the safe load that can be applied without damaging the pipe. Geometry techniques such as laser scanners or light lines are also used to determine changes or deformations in transverse sections and along the pipe.

In order to detect leaks, flow monitoring weirs, smoke/gas testing and sonic leak detector methods can be used in combination with CCTV and the methods described above.

2.3.2 Assessing Hydraulic Condition

The hydraulic condition of a sewer system can be assessed mainly in two ways: one way is to use the theoretical loading factor (TLF) and the other is the so-called grade line factor (GLF).

The theoretical loading factor is the ratio between the peak flow and the pipe capacity. This indicator gives an indication of undersized sewers. However the indicator is not able to reflect or measure where flooding will occur.

The grade line factor uses the maximum hydraulic grade line (HGL) reached during an extreme or design event relative to the pipe crown, basement elevation and ground level. Using this information it is possible to estimate surcharge and flow restriction in the system, indicating the flooding risk. The determination of the HGL requires the use of models. Their use is very important when assessing hydraulic conditions in sewers and it is treated in more detail below. The Federation of Canadian Municipalities uses Table 5 for the hydraulic condition performance evaluation of sewers.

Depuis (2000) proposed a method for evaluating the hydraulic performance of a sewer network, based on the comparison between the HGL (hydraulic gradient slope) and the invert slope. The greater the ratio of HGL slope to sewer invert slope the higher the need for hydraulic rehabilitation. However, this methodology is not a sufficient condition to identify problems upstream caused by the analyzed section or backwater effects. Important factors such as topography, location of the pipes, vulnerability of the site and overflows must also be taken into account.

Surcharge and flooding due to backwater effects are the main hydraulic parameters to take into account for a rehabilitation process. When the maximum TLF ratio is less than unity a free surface (non pressure) condition is assumed. When the capacity of the pipe is exceeded (TLF more than unity) surcharge is caused, the flow regime passing from free surface to pressure flow (Reyna et al 1994).

Table 5: Hydraulics condition ratings

TFL	Upstream Impact	Hydraulic Condition Rating
Less than 1.0	No impact on upstream GLF	1.0
1.0 – 2.5	Upstream GLFs are generally between 1 and 2. Pipe link contribute to GLF	2.0
1.0 – 2.5	Upstream GLFs are generally between 2 and 3. Pipe link contribute to GLF	3.0
2.5 and greater	Upstream GLFs are generally between 3 and 4. Pipe link contribute to GLF. Potential exists for basement flooding.	4.0
2.5 and greater	Upstream GLFs are generally between 4 and 5. Pipe link contribute to GLF. Potential exists for basement flooding.	5.0

In order to evaluate the hydraulic performance the following criteria can be analyzed alone or in combination:

- Hydraulic capacity of the network,
- Frequency and duration of surcharges,
- Overflow volume,
- Duration of flooding,
- Spatial area affected by flooding and
- Cost of damage to public or private property

The previously enumerated criteria are symptoms and they can be the product of several causes which promote changes in the hydraulic grade line such as pipe diameters that are insufficient to drain the runoff volume, the decrease of slope due to damaged bedding, the increase of roughness as a product of aging or sedimentation, the presence of blockages causing minor losses, and the increase of the tail water level at the discharge point, among others.

The hydraulic failure of a pipe can be defined using the value of a hydraulic performance parameter (H^{US}), see Figure 2.1. When H^{US} is equal to zero the pipe is working under free surface flow conditions and it can be said to have an adequate hydraulic condition. On the other hand, when H^{US} reaches the ground surface or house basement level flooding is present. Bennis et al (2003) propose a hydraulic performance index for sewer networks based in these concepts. They propose the use of a linear equation that depends on the ratio between H^{US} and the depth of the pipe crown from the ground surface (G) expressed as a percentage (Eq. 1):

$$N = 100 \times \frac{H^{US}}{G} \qquad (1)$$

N varies in a range between N_{MIN} = 0% and N_{MAX} =100%. When H^{US} is equal to G this means that the water level has reached the ground surface and flooding is likely. In this case N=100% (N_{MAX}). The other extreme is when N=0% (N_{MIN}); for this case the flow is not under pressure so free surface flow is present and can be said that the pipe is adequate under the current demands so it does not require changing.

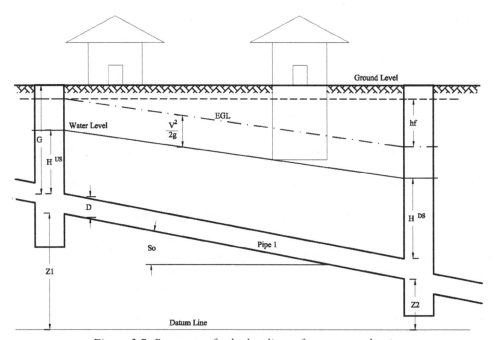

Figure 2.7: Parameter for hydraulic performance evaluation

Eq. 1 is the most basic case. Some local design criteria may require a maximum value of H^{US} in order to avoid reaching the ground level. This can be achieved by setting a minimum and maximum value for N. Also N is not linear as in Eq. 1. The surcharge at levels near of the pipe crown are less important than when the surcharge almost reaches the ground level.

In other words, if H^{US} is close to the pipe crown the situation is less dangerous than when H^{US} is very close to the ground or the basement of a house. So Eq. 2 has been proposed to take into account these last two issues:

$$N = N_{MIN} + (N_{MAX} - N_{MIN}) \times \left[1 - \left(1 - \frac{H^{US}}{G} \right)^n \right]$$
(2)

In Eq. 2, n allows us to introduce a nonlinear effect. It represents the vulnerability of the area to be drained by the pipe. Areas where flooding has no consequence a value of n equal to zero can be used. In the other hand, areas which are highly vulnerable to floods are expected to have values of n of two or three to reflect the non linearity of the equation for N. If a linear expression is accepted a value of one for n is used.

An undersized pipe downstream may contribute to the surcharge in a pipe located upstream. The surcharge or the increase of the water level in a pipe section can raise the HGL and produce several surcharged pipes upstream of such a section. This is more significant when sewer slopes are not steep. In order to include such a backwater effect Eq. 2 has to be adjusted. If the Bernoulli equation is applied between manholes upstream and downstream, it can be inferred that H upstream (H^{UP}) value has to be corrected because the H downstream (H^{DS}) promoted by the section downstream as is shown in Eq. 3:

$$N = N_{MIN} + (N_{MAX} - N_{MIN}) \times \left[1 - \left(1 - \frac{H^{US} - H^{DS}}{G} \right)^n \right]$$
(3)

There may be other causes for pipe surcharge like blockages, root intrusion, sedimentation, etc,. However, such failures can be corrected using the normal maintenance procedures. The present research is focused in the major rehabilitation work of a sewer system.

2.3.3 Assessing Environmental and Social Conditions

As sewers age a sewer rehabilitation work plan becomes mandatory not only for maintaining the sewer system at the current levels of service but also for avoiding the alleviation of potential adverse environmental and social impacts. The main environmental impact is the overflow discharges to the receiving water body. If the pollutant mass or concentration is high it can produce severe damages to aquatic life or users downstream. Another source of pollution is the ex-filtration of sewage to aquifers. Less considered problems but none the less important are diseases promoted by vectors proliferation, bad odors and poor aesthetics due to litter.

Environmental impacts are assessed using water quality models. In this case Dissolved Oxygen, Ammonia and Suspended Solids may be predicted directly by the combined use of a water quality sewer model and a water quality river model. Some of the indicators are mentioned in previous sections.

The assessment of the ex-filtration to aquifers is done through performance indicators. The estimation of such performance indicators requires flow measurements in the sewer network. Three main indicators can be defined for water ex-filtration in sewers (Cardoso A et al. 2006) (Bertrand-Krajewski J-L et al. 2006)(Equations 4,5,6).

$$I_1 = \frac{Q_{exf}}{Q_{avedwf}} \tag{4}$$

$$I_2 = \frac{Q_{exf}}{L_{sewer}} \tag{5}$$

$$I_3 = \frac{Q_{exf}}{A_{sewer}} \tag{6}$$

where: Q_{exf} is the ex-filtration flow

Q_{avedwf} is the rainfall intensity

L_{sewer} is the length of sewer

A_{sewer} is the plain area of the sewer (L*A)

In Eq. 4 the ex-filtration flow is expressed as a percentage of the daily mean dry weather flow. A drawback of this indicator is its dependency on the values of the dry weather flow. Amick and Burgess (2000) present values of experimental studies with values of I between 11.9% and 49%. Ellis et al. (2002) comment that probably this value of I is not greater than 10%. Eq. 5 represents the mean ex-filtration flow per unit length of sewer. This indicator gives relevant results in systems where ex-filtration takes place predominantly along the sewers. Literature report values for ex-filtration between 0.000027 and 0.001 m^3/day/m (Amick and Burgess, 2000). Finally, Eq. 6, gives the mean ex-filtration flow per unit area of sewer wall. This indicator is related with the sewer wall area that is potentially subject to ex-filtration. Sewer longitudinal area is calculated as the sum, for all sewers of the measured reach, of the pipe perimeter times pipe length. Literature values are between 0.08 and 1.20 m^3/day/(cm.km).

Other causes of environmental impact such as odors are often easily determined by simple analysis and by feedback from the general population. To assess the environmental impact of this type requires a rigorous assessment of operational and maintenance tasks, feedback from operators and the evaluation of concerns expressed by the public.

2.4 Rehabilitation Options

As mentioned above, rehabilitation techniques involve the correction of the performance of the network in terms of environmental, structural and hydraulic defects. The cost of each alternative depends on the employed technique and materials. Abraham and Gillani (1999) present a discussion about some rehabilitation techniques and when one is to be recommended over an other. The most common techniques can be identified as:

- Excavation and replacement (renewal)

- Chemical grouting, Pipe lining, Coatings

- Storage and real time control (RTC)

- Diversions

Excavation and replacement are carried out when a structure has collapsed or when there is a large crack or additional capacity is needed. The main problem with such a technique is the interruption to the service, so temporary solutions must be provided in order to maintain the flow of the sewage. Also disruption of normal activities in the vicinity of the repair is of considerable importance. In particular, the impact on the traffic and the environment must be taken into account.

Chemical grouting is employed to avoid infiltration from groundwater to the system, or leakage from pipes. It is aimed at closing cracks in pipes or to filling voids around them. It is a cost-effective technique in rehabilitation. Common types of chemical grout are acrylic grout, urethane grout and urethane foam.

Pipe lining is the process of inserting a lining inside the pipe aimed at restoring its performance. These methods can reconstruct the pipe without disrupting traffic. The methods of lining are:

- Slip-lining which consists in sliding a flexible pipe of slightly smaller diameter into the existing pipeline. It can be made of high-density polyethylene, fiberglass-reinforced plastic, un-plasticized polyvinyl chloride, etc.

- Cured-in-place pipe lining (Inversion lining) formed by inserting a resin impregnated felt tube into the existing pipe. The tube is inverted against the inner wall of the pipe and left to cure.

- Deformed pipes are repaired by inserting a thermoplastic polyethylene or polyvinyl chloride pipe into the existing pipe, then expanding the insert to form a tight fit with the sewer. This method needs less time to cure.

- Segmental lines consists of connecting segments (around 100 cm) of pipe inside the existing pipe. It is adopted when the use of flexible pipes is not possible. However, this process is labour expensive.

Coating is used to avoid abrasion and corrosion in existing pipes. It also extends the lifetime of the pipes. It is estimated that coating increases lifetime in concrete, steel and iron pipes by up to 30-50 years. Several types of coating are in common use, including concrete mortar, reinforced gunite and resin coatings.

Robotic rehabilitation in one of the most versatile modes of extending the lifetime of pipe networks. These systems are operated by hydraulic motors and allow the repair of inaccessible parts of the network. This technique falls between the lining and chemical approaches.

Storage is used as good practice. This permits some areas such as parks to flood or relies on artificial storage constructed to reduce the size of pipes in the underground network. This type of solution is implemented when it is not possible to increase the diameter of a pipe or it is too expensive.

Knowledge of these techniques helps to build a cost function in order to evaluate the performance of the network.

2.5 Sustainable Urban Drainage

The term of "sustainability" was introduced in the 1990's in the UN Earth Summit at Rio de Janeiro in 1992 and it has had an important influence on the urban drainage design and rehabilitation paradigm. The balance between economical, technical, environmental and social issues becomes of major importance. The sewerage assets represent not only a guarantee, saving people's belongings and lives from floods,but they must provide a better quality of life in a more wider sense. Aesthetic, health, and recreation benefits have to be included in any rehabilitation plan.

Approaches in drainage management involve a variety of stakeholders in the selection of a sustainable drainage system; the design of a rehabilitation plan needs to consider not only water quantity but also water quality and amenity (Ellis et al, 2004). It must integrate in a balanced way technical, environmental, economical and social interest (Figure 2.8). Sustainable urban drainage system (SUDS) approaches are aimed at including social and environmental factors in the final decision making.

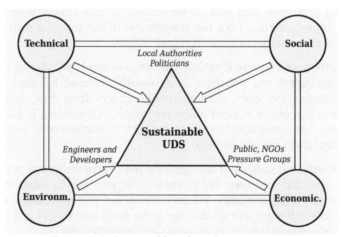

Figure 2.8: Sustainable urban drainage system

Traditional urban drainage has a short term vision: drainage systems are designed or rehabilitated to evacuate excess water as soon as possible, and where the economical aspects have bigger weighting factors than social or environmental aspects. This does not take into account cumulative damages over long periods of times. The use of SUDS techniques , over longer time horizons, reduces long term damages, and environmental or even social costs are given more importance. The main aim of these systems is to enable better management of runoff (in terms of both quantity and quality) at source by preserving natural drainage patterns and emulating the natural hydrological cycle. Examples of these measures are: runoff reuse, soakaways, infiltration trenches, swales, green roofs, permeable paving, ponds, wetlands, etc. (see also Vojinovic and Abbott, 2012).

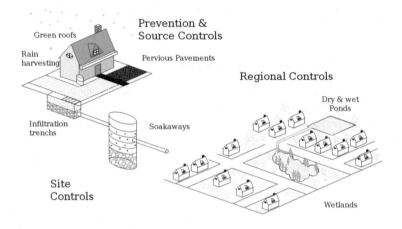

Figure 2.9: SUDS prevention and control measures

The main idea is to return excess water into final disposal as close as possible from the source, avoiding in this way high peaks in the network system. Several stages are proposed to fulfill that purpose, starting from the management of the source, runoff and pollutant prevention, to different stages of controls (Figure 2.7):

- **Prevention and Source Control**: This stage is aimed at reducing the quantity of the total runoff and pollutants. It is usually achieved by implementing good housekeeping measures, cleaning runoff surfaces from dust, greases, and any pollutant that can be washed off by the rain. Rain harvesting is also a prevention measure. Some non structural measures such as implementing particular policies and regulations can also be used.

- **Site Control**: It consists of managing the runoff near the source or on site through some structural measures. For example infiltration cab be increased through the use of pervious pavements; infiltration trenches and soakaways help to return water in the groundwater system; and green roofs retain water and then transfer it to the atmosphere through evaporation and evapotranspiration.

- **Regional Control**: The runoff and pollutants collected from several sites are conveyed to regional control measures before their final disposal in order to attenuate peak discharges and reduce the need for treatment. This is achieved by incorporating dry or wet ponds, wetlands, storage, etc, in the drainage system.

The use of the above cascade of controls, ensures a reduction of peak flows and pollution to reaching the main sewer system. It also promotes the co-responsibility of stakeholders in the integral management of the system.

2.6 Urban Drainage Modeling

This section is aimed at describing the present state-of-the-art in urban storm drainage modeling. Most of the problems related to flooding in urban zones are due to insufficient capacity of the sewer systems and intake structures (Russo 2005). This can be due to several factors, such as the fast growth of urban areas (Mark 2004), poor designs, and climate change impacts (Wright 2005). Also sediment and floating objects can partially occlude underground conduits, diminishing their capacity (Gupta 2005).

Mathematical models play a major role in designing and improving storm water collection systems. Simulation results are widely used for planning, design, rehabilitation and operational purposes   (Ji, 1998). Computational hydraulic models are used as virtual laboratories to mimic the real sewer network; scenarios and alternatives can be evaluated in order to make the best investment choice. The selection of an appropriate hydraulic model is of considerable importance. . The mathematical model of sewer system can be as simple as a formula or as complex as a solution of the complete 3D hydrodynamic equations (see for example, Price and Vojinovic 2011 ; Vojinovic and Abbott, 2012).

2.6.1 Storm-water System components

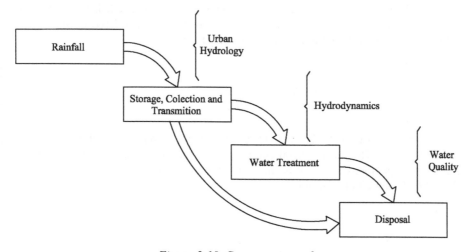

Figure 2.10: Sewer system schema

A storm-water system can be schematized as shown in Figure 2.10. We can distinguish the input which is the rainfall (hydrology), the sewerage components or conduction system (hydraulics), the treatment system (quality), and final disposal.

In order to build a model of a sewer system, it is necessary to know the components of such a system. An approach to identify consistent components and to model them separately is used to have a better understanding of the processes and components. The main division is to separate the hydrological processes from the hydraulic processes.

2.6.2 Rainfall-runoff modeling

Design rainfall

The forces which drive the complete drainage process are the rainfall and waste-water inputs. There are two main kind of rainfall that are used to feed a model, depending on the type of analysis that is needed. The rainfall inputs can be a single event, usually associated with a return period or probability of occurrence, or a historical time series. The most common input is a single event associated with a frequency of occurrence or the so called return period, duration and rainfall depth or intensity.

The return period can be defined as a function of the frequency:

$$T_R = \frac{1}{P(X \geq x)} \tag{7}$$

where T_R is the return period, and $P(X>x)$ is the probability that a given X event is exceeded in a given period of time. For instance if an event X is exceeded 2 times in 10 years it corresponds to a probability of 1 in 5. In other words it has a 5 year return period.

In order to find $P(X=>x)$ a record of maximum annual rainfall is needed. If n is the total number of years with data and m is the rank of a value in a list ordered by descending magnitude, the exceedence probability of x_m for a large n is:

$$P(X \geq x) = \frac{m}{n} \tag{8}$$

Eq. 8 is known as California's formula. There are several empirical distributions used to calculated $P(X=>x)$. The most general empirical distribution can be written as shown in Eq 9 . In such a equation, if $b=0.5$ it is known as Hazen's formula, $b=0.3$ as Chegodayev's formula and $b=0$ as Weibul's formula.

$$P(X \geq x) = \frac{m-b}{n+1-2b}$$
                                                                                    (9

Intensity, Duration and Frequency relationship (IDF Curves)

An useful source of information are IDF curves. IDF curves belongs to a specific location at which rainfall data was collected. For instance Figure 2.11 shows typical IDF curves for Cabudare city in Venezuela. It can be seen how the intensity decreases with the duration of the rainfall and increases when the return period is increased.

Building IDF curves is done using so-called frequency analysis. The procedure consists in ranking by descending order the maximum annual rainfall grouped by duration, and then making an estimation of the exceedence probability using an empirical equation like Eq. 9. Finally, the probability and the corresponding intensity are fitted to a probabilistic statistical distribution. There are several probabilistic distribution that are commonly used in hydrology; see Table 6. However, the most commonly used distributions in urban hydrology for rainfall are the extreme value functions, and the more specific Extreme Value Type I (Gumbel distribution when $u=0$ and $\alpha=1$) for design rainfall.

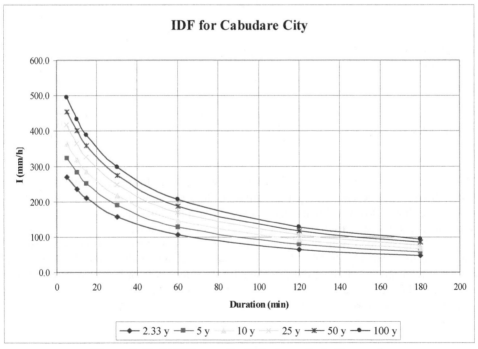

Figure 2.11: IDF Curves for Cabudare city in Venezuela

Table 6: Probability distribution functions

Distribution	Equation	Parameters
Normal	$f(x)=\dfrac{1}{\sigma\sqrt{2\pi}}\exp\left[(-\dfrac{(x-\mu)^2}{2\sigma^2})\right]$	$\mu=\bar{x},\ \ \sigma=s_x$
Log normal	$f(x)=\dfrac{1}{x\sigma\sqrt{2\pi}}\exp\left[-\dfrac{(y-\mu_y)^2}{2\sigma_y^2}\right]$	$y=\log(x)$ $\mu_y=\bar{y},\ \ \sigma_y=s_y$
Exponential	$f(x)=\lambda e^{(-\lambda x)}$	$\lambda=\dfrac{1}{x}$
Gamma	$f(x)=\dfrac{\lambda^{\beta}x^{(\beta-1)}e^{-\lambda x}}{\Gamma(\beta)}$	$\lambda=\dfrac{\bar{x}}{S_x^2}\ \ \beta=\dfrac{\bar{x}}{S_x^2}$
Pearson Type III	$f(x)=\dfrac{\lambda^{\beta}(x-\epsilon)^{\beta-1}e^{-\lambda(x-\epsilon)}}{\Gamma(\beta)}$	$\lambda=\dfrac{S_x}{\sqrt{\beta}}\ \ \beta=\left(\dfrac{2}{C_s}\right)^2$ $\epsilon=\bar{x}-S_x\sqrt{\beta}$
Log Pearson Type III	$f(x)=\dfrac{\lambda^{\beta}(y-\epsilon)^{\beta-1}e^{-\lambda(y-\epsilon)}}{y\Gamma(\beta)}$ *where* $y=\log x$	$\lambda=\dfrac{S_y}{\sqrt{\beta}}\ \ \beta=\left(\dfrac{2}{C_s(y)}\right)^2$ $\epsilon=\bar{y}-S_y\sqrt{\beta}$
Extreme Value Type I	$f(x)=\dfrac{1}{\alpha}\exp\left[-\dfrac{x-u}{\alpha}-\exp\left(-\dfrac{x-u}{\alpha}\right)\right]$	$\alpha=\dfrac{\sqrt{6}\,s_x}{\pi}$ $u=\bar{x}-0.5772\,\alpha$

This procedure must be repeated for as many durations of rainfall as are available. In other words there would be one equation per duration. It is a common practice to use existing curves like those shown in Figure 2.11. Also, when a storm-water system is designed or rehabilitated, it is necessary to use different values of duration that may or may not match with the data provided from the gauges. The most common durations from recorded data correspond to 5, 10, 15, 30, 60, 120, 180, 360, 540, 720 and 1440 minutes. For practical reasons is advisable to fit these durations against the intensity in order to have a continuous equation in duration for a fixed return period. There exist many equations that can fit this type of data but the most accepted is:

$$i=\frac{a}{(D+b)^c} \qquad (10)$$

where $a,b,$ and c are constant values, i is the intensity in mm/h, and D is the duration of the design rainfall. One equation per return period is used.

Design Rainfall Duration

The duration of the rainfall is a key-parameter for the rehabilitation or design of drainage systems. It is expected that the peak discharge occurs when the rain drop that falls farthest from the outlet leaves the catchment. The time required for this particle to travel from the catchment divide to the catchment outlet is called the concentration time. In other words, it is expected that design rainfall duration has to be at least equal to the concentration time. The concentration time (t_c) can be divided in two main parts: the travel time overland ($t_{overland}$) (sometimes called the "*time of entry*") and the travel time in the channels ($t_{channel}$) (Eq. 11).

$$t_c = t_{overland} + t_{channel} \qquad (11)$$

There is no analytical procedure to compute the concentration time. The travel time depends on the rainfall intensity, and the design rainfall depends on the travel time. Most equations for t_c therefore are empirical, and are developed or estimated under specific conditions and for specific locations. However, they have been applied successfully elsewhere as well. In urban drainage the following equations have a generalized use:

- Kirpich (1940): developed by the Soil Conservation Service (SCS). This equation was developed for rural watersheds in Tennessee (USA).It includes the travel time overland and in channels. In order to be applied to urban areas a correction factor G of 0.4 must used for surface flow over concrete or asphalt. A correction factor of 0.2 must be applied for travel time in concrete channels. No adjustment is needed for overland flow on bare soil or road side ditches. Eq. 12 shows Kirpich's concentration time equation.

$$t_c = 0.01947\, G\left(\frac{L}{\sqrt{S}}\right)^{0.77} \qquad (12)$$

  where t_c is the concentration time in minutes, L is the length of the channel or ditch in meters from the highest place to outlet and S is the average catchment slope.

- California Culverts Practice (1942): This is basically the same as Kirpich's equation but is adapted to California's small mountain basins.

$$t_c = 0.01949\, G\left(\frac{L^3}{H}\right)^{0.385} \qquad (13)$$

  where H is the elevation difference between the catchment divide and the outlet in meters.

- Izzard (1946): the Bureau of Public roads (USA) developed his equation in the laboratory for the overland flow on roadways.

$$t_c = \frac{3.1952\,(0.01778\,i+c)\,L^{1/3}}{S^{1/3}\,i^{2/3}} \tag{14}$$

where i is the rainfall intensity in mm/h, c a retardance coefficient, L length of flow path in meters, and S the slope of the flow path. The retardance coefficient varies in a range between 0.0070 for very smooth roads to 0.06 for dense turf. The solution of Eq 14 requires the use of iteration. An important observation is that product i times L must be less than 645 m.

- Federal Aviation (1970): This equation was developed to be used on airfields by The US Corps of Engineers. However it has been used for overland flow in urban areas. One of the advantages is that it uses the same C coefficient for the rational method to compute runoff.

$$t_c = 1.8(1.1-C)\frac{L^{0.5}}{S^{1/3}} \tag{15}$$

where L is the length of overland flow in ft and S is the surface slope in percentage.

- Kinematic Wave: It was introduced by Linsley in 1965 and improved by Aron and Eborge in 1973. It was developed using a kinematic wave analysis on surfaces.

$$t_c = \frac{0.94\,L^{0.6}\,n^{0.6}}{i^{0.4}\,S^{0.3}} \tag{16}$$

where L is the overland flow length (ft), n the manning roughness coefficient, i the rainfall intensity in in/h, and S is the average overland slope.

- SCS lag equation (1973): This was developed initially using rural agricultural areas by SCS. However it has been adapted to small urban catchments with areas less than 800 ha. It has proven to perform well in paved areas but over estimate on mixed areas. (chow, et al 1988).

$$t_c = \frac{100\,L^{0.8}\,[(1000/CN)-9]^{0.7}}{1900\,S^{0.5}} \tag{17}$$

where L is the hydraulic length of watershed, CN is the SCS runoff curve number and S is the average watershed slope in percentage.

Areal Reduction Factor

The depth computed using IDF curves, time of concentration (duration) and the selected return period has to be corrected by an areal factor. The measured values on the gauge station are at a point; this means they are valid for only a few square kilometers from the storm center and the average rainfall depth over an area is less than the point precipitation. The ratio between point rainfall depth and the average rainfall over an area is called the Areal Reduction Factor (ARF).

The two types of areal reduction factors commonly in use are Geographically or Area Fixed and Storm Centered relationships. The fixed area ARF is based on the frequency analysis of a time series of annual maximum precipitation for a fixed area; an assumption is made about the equality of the probability distribution function of the point and the areal rainfall. The second type corresponds to any given real storm; it represents the ratio between the average rainfall over an area and the center of the storm. Storm centered ARFs are used in probable maximum flood estimation, while the fixed area ARFs are used for designing hydraulic structures for flood control (Sivapalan and Blöschl 1998).

Areal reduction factors have been mostly developed in the U.S., U.K. and New Zealand (Omolayo 1993). Not much work has been done to estimate these values in other parts of the world because of the sparse networks of rainfall stations and short records. There are many methodologies for the transposing these areal reduction factors to different parts of the world (Gill 2005). Details of some of the major methodologies in practice for the derivation and transposition of areal reduction factors can be found in Rodriguez-Iturbe and Mejia (1974), Gill (2005), Butler and Davies (2004).

In the UK and other European countries the *"Wallingford"* procedure for ARF is used. Eq 18 shows the general form of the equation. In such an equation f_1 and f_2 are regional parameters that can be empirically estimated depending of the region or local conditions and D is the storm duration. In The Netherlands statistical ARF have been determined using similar equations (Witter 1983).

$$ARF = 1 - f_1 D^{-f_2} \tag{18}$$

Storm-centered methods, such as some correlation-based methods and the so-called annual maxima-centered method, do not result in probabilistically correct areal rainfall estimates. That is, when multiplying a storm-centered ARF with a T-year point rainfall, the resulting areal rainfall does not necessarily have the same T-year return period. These methods can therefore not be recommended for use with rainfall frequency estimates. Rainfall radars can be used to estimate ARF curves. However, the use of radar data is problematic. Radar has also been used to determine ARF curves; an comparison between small and large radar image use can be found in Allen and DeGaetano (2005)

Rainfall Temporal Distribution

In Addition to the spatial distribution of the rainfall, considered in the ARF curves, it is necessary to distribute the rainfall also in time. In order to distribute the design rainfall a dimensionless or synthetic hyetograph is needed. A design hyetograph is the time distribution of the total rainfall depth corresponding to a duration and return period and corrected by the ARF factor. Dimensionless hyetographs are measured rainfall events that has been reduced or normalized to be dimensionless using the rainfall duration and rainfall depth at any time (Nascimento et al. 2000).

The most common and wide spread distribution is a rectangular distribution, maybe because this distribution is effectively used by the rational formula and because of its simplicity (Pilgrim and Cordery, 1993). More complex methods have been implemented, like the triangular hyetograph (Yen and V. T. Chow 1980), the Watt hyetograph (1986), the Desborde hyetograph and the "Best Linear Unbiased Estimation" (BLUE) hyetograph (Alfieri et al. 2008). Recently, Grimaldi and Serinaldi (2006) develped a multivariate approach and Lin et al. (2009) proposed a method using self organizing maps (SOM) for the regionalization of design rainfall hyetographs.

Other important groups of papers published on rainfall distributions came from research at the Illinois State Water Survey. Huff (1967) analyzed rainfall data from heavy rainstorms collected during the period from 1955 through 1966. He observed a major portion of the total storm rainfall occurs in a small part of the total storm time, regardless of storm duration, areal mean rainfall, and the total number of showers or bursts in the storm period. The Soil Conservation Service (1986) method is a variation of the Huff method. The country (USA) was divided into two geographic regions which are represented by the SCS Type-I and Type-II design rainfall distributions. The distributions are based on the depth duration curves for a return interval of 25 years.

However, one of the more used methods in engineering practice is the Chicago method, introduced by Keifer and Chu in (1957) . The method is based on Eq. 10, which represents an average intensity in order to estimate the instantaneous intensity for a given time. Eq 10 is multiplied by the duration t of the rainfall to obtain the rainfall depth P for a given time. Then the result of the multiplication is derived in order to obtain the instantaneous intensity.

$$i = \frac{dP}{dt} = \frac{a[(1-c)t+b]}{(t+b)^{c+1}} \tag{19}$$

Eq. 19 represents only half of the real intensity. However, for a real storm the intensity starts with low values, increases around the middle of the duration and then returns to low intensity values again. For this reason the total duration t is divided in two parts: t_a and t_b. The relation between t_a and t_b is given by Eq. 20, where r is a ratio.

$$t = \frac{tb}{r} = \frac{ta}{1-r} \tag{20}$$

Using the Eq. 19 and 20 two further equation are obtained for the instantaneous intensity. One equation is for a time interval between $t=0$ to $t=t_a$ and the other for an interval from $t=t_a$ to $t= t_b$. (Eq. 21). Figure 2.12 shows the final instantaneous graph for the total rainfall event duration. The integration of intensity equation must be equal to the rainfall depth event.

$$i_b = \frac{a[(1-c)t_b/r+b]}{(t_b/r+b)^{c+1}}$$

(21)

$$i_a = \frac{a[(1-c)t_a/(1-r)+b]}{(t_a/(1-r)+b)^{c+1}}$$

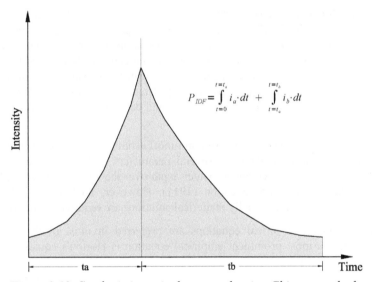

$$P_{IDF} = \int_{t=0}^{t=t_a} i_a \cdot dt + \int_{t=t_a}^{t=t_b} i_b \cdot dt$$

Figure 2.12: Synthetic intensity hyetograph using Chicago method.

Prodanovic and Simonovic (2004) carried out an investigation into a synthetic rainfall hyetograph for the city of London. They concluded that the method of Watt et al. (1986) should probably not be used in cases where storm durations of 24 hr and longer are needed. This is because the derived parameters in the method are relevant only to 1-hr storms. However, the method of Yen and Chow (1980) does not suffer these drawbacks, and should therefore be used, simply because it allows for a wider range of storm durations. if hydrologic extremes are to be investigated (which may exceed 100 yr return period), the method of Keifer and Chu (1957) should also not be used, unless data for longer return periods are obtained. The SCS (1986) method should be used when rainfall of high intensity within a short period of time is needed. The Huff method can be used when rainfall of low intensity, distributed over a longer period of time within the same storm duration, is required.

Hyetographs can be historical. Because a historical hyetograph represents only a single sample from the temporal distribution population, it may not be representative of the whole event. Therefore, designers prefer the use of synthetic design hyetographs. Another alternative is to use continuous data. However, the use of continuous rainfall data may require high computational power and large time series.

Runoff Generation

After the design rainfall is estimated and spatially and temporal corrected, the next step is to transform of the rainfall into runoff. The volume of the precipitation is divided into two main parts: The rainfall losses, and the proportion of rainfall that runs over the surface (runoff). The losses can be due to multiple factors like rainfall interception, storage in surface depressions, evaporation and infiltration.

Before the rainfall reaches the soil surface, it can be intercepted by the tree canopy or other objects such as house roofs. These objects have the capacity to store or retain part of the rain water such that it never reaches the soil or the drainage system. This water can only leave the system through an evaporation process. Water can also be retained by depressions in the surface, in the same way than interception water, the water in depressions never reaches the drainage network. The water is evaporated but can also be infiltrated into the soil. It is common practice to merge interception and depression storage processes into one common process.

The infiltration process plays a critical role in runoff estimation and it is the main source of rainfall losses. Infiltration depends on several parameters such as soil conductivity, soil structure, soil moisture, surface cover, water depth over the soil, and others. Infiltration is described by the Green Ampt equation (1911). However, analytical solutions of this equation are not common and usually numerical solutions are required (Serrano 2003).

For the reason above, empirical equations are preferred in most of the rainfall-runoff models. May be the most prominent empirical equation is Horton's equation (1933) (Eq. 22).

$$f(t) = f_c + (f_o - f_c)e^{-kt} \tag{22}$$

where: $f(t)$ is the infiltration rate at time t

f_o, f_c is initial and final infiltration rate

k is a decay constant

In urban hydrology other simpler equations are used, depending on the design method. If the time distribution of the rainfall is not considered, a constant value for the losses is sufficient. If time is considered, more complex methodologies are required. The most common methods are discussed next.

Rational Equation

The rational method was originally developed to estimate the peak runoff. It uses as input the intensity provided by IDFs curves and does not take into account the temporal rainfall distribution. The losses are assumed to be constant and are taking into account using a constant factor C. Eq. 23 shows the rational formula.

$$Q = C \cdot I \cdot A \qquad (23)$$

where: Q is the peak discharge

               I is the rainfall intensity

               A Is the area of the catchment

The rational formula has been used since about 1889. It is appropriate for estimating peak discharges for small drainage areas of up to about 80 hectares in which no significant flood storage appears. The runoff coefficient (C) represents the integrated effects of infiltration, evaporation, retention, flow routing, and interception, all of which affect the time distribution and peak rate of runoff. The factor C depends on the land use and values can be found in the literature; see Chow et al. (1988) Butler and Davis (2004) propose in residential areas the values ranging from 0.5 to 0.9 depending of the land use type. When the drainage area has a heterogeneous land use, it is common practice to make a weighted average over the area.

A modification of the rational method known as "Wallingford" method was introduced in which the C coefficient is composed by two coefficients: $C = C_V \cdot C_R$, where C_V is a volumetric coefficient and C_R is a coefficient to take into account the routing.

The use of the rational formula is subject to several limitations and procedural issues in its use:

- The most important limitation is that the only output from the method is a peak discharge (the method provides only an estimate of a single point on the runoff hydrograph).

- The simplest application of the method permits and requires subjective judgment by the user in its application. Therefore, the results are difficult to replicate.

- The average rainfall intensities used in the formula have no time sequence relation to the actual rainfall pattern during the storm.

- The computation of t_c should include the overland flow time, plus the time of flow in open and/or closed channels to the point of design.

- The runoff coefficient, C, is usually estimated from a table of values. The user must use good judgment when evaluating the land use in the drainage area under consideration.

- Many users assume the entire drainage area is the value to be entered in the Rational method equation. In some cases, the runoff from only the interconnected impervious area yields the larger peak flow rate.

Time Area Method

The time area method was introduced around the 1920s and its early use was principally in channel routing procedures (Calver 1993). In its simplest form, the time-area method computes runoff from a catchment as a sum of time-distributed rainfall values falling on spatial divisions of a catchment defined by assigning a constant velocity to the runoff.

The method is based on the rational method and computes the losses using the C coefficient. However, it is able to manage temporal rainfall distribution. The rational method does not take into account the shape of the catchments. This issue is overcome with the time-area method and it can be demonstrated that the shape of the catchment has an influence on the output hydrograph. However, this method like the rational formula tends to overestimate the peak discharge.

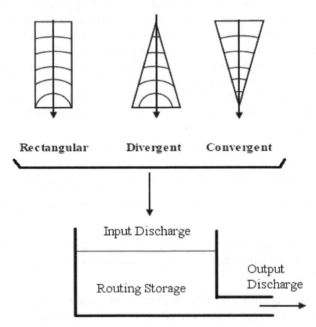

Figure 2.13: Time-area method schematization

The Transport and Road Research Laboratory (TRRL in UK) method is a variant of the time-area method developed by Watkins and Young in 1962-1965 and improved for tropical countries by Watkins (1976). It is assumed that only the impervious areas are connected to the drainage system and contribute to the peak flow. Discharges are calculated from a time-area diagram and then modified by a reservoir routing equation (Colyer 1977) as shown in Figure 2.13. The TRRL method, and/or modification of it, is included in most of the actual urban modeling tools in order to estimate the runoff hydrograph. The method is simple to apply, additional parameters are easily included, and the formulation has a basis of physical reasoning (Calver 1993).

SCS Curve Number Method

The SCS Curve Number method is a more complex method. It is an empirical method developed by the US Natural Resources Conservation Service (NRCS), former US Soil Conservation Service (SCS). The project started around 1930s and 1940s with the installation of multiple infiltrometers to measure soil infiltration in several US rural watersheds. From this data an empirical equation was derived to determine the runoff volume from rainfall volume. The method was developed in 1956 and it has been revised several times, the latest in 1993 (Surendra Kumar Mishra and Singh 2003).

Although the equation was originally developed for the design of structures in rural areas, it has been used in urban areas as well. The estimation of the runoff is carried out for a single rainfall event and has become a standard in the US (NRCS 1986). One of the advantages is that the methodology is very simple in only relying on one parameter, the so called curve number (CN). Eq. 24 shows the relationship between the input rainfall volume without initial abstractions, (interception and storage losses Pe), and the runoff (Q).

$$\frac{F}{S}=\frac{Q}{P_e} \tag{24}$$

where: Q is the runoff volume

Pe is the effective precipitation (without initial abstraction)

F is the infiltration volume

S is the potential maximum retention after runoff begin

From the measured data the initial abstractions (Ia) were estimated like $Ia=0.2S$, and by continuity equation $F=Pe-Q$ leading to following equation:

$$Q=\frac{(P-0.2\cdot S)^2}{(P-0.8\cdot S)} \tag{25}$$

Also, from measured infiltration data the value of S has been estimated as shown in Eq. 25. As can be seen, the potential maximum storage (S) is a function of the curve number (CN). The curve number selection is done taking into account the variables: land use cover, hydrological soil type and the soil moisture content when the rainfall events occurs.

$$S=\frac{1000}{CN}-10 \tag{26}$$

Land use ranges from forest zones to impermeable urban zones. The soil type is divided basically into four types: Type A with the lowest runoff potential (gravel), Type B with moderate runoff potential (sandy soils), Type C with moderate to high runoff potential (shallow clay soils) and Type D with a high runoff potential (swelling clays). The CN can be estimated from literature values; see NRCS (1986) and Mishra and Singh (2003) for more references.

A critical parameter is the soil moisture at the moment of the rainfall. In SCS the method is defined as a soil hydrological condition. The original method was developed to be used for one event. The introduction of a modification was carried out in order to take into account soil moisture (Hawkins 1978). Antecedent moisture conditions (AMC) is taken into account by changing the value of the curve number. The original tables in the literature give the average condition (AMC-II) and table 7 shows the conditions for AMC-I and AMC-III. This change has been introduced to apply the method also for a continuos rainfall time series (Geetha et al. 2007).

Table 7: Antecedent moisture conditions (AMC)

AMC	Total five-days antecedent rain fall (cm)		
	Dormant season	Growing season	Equation
I	Less than 1.3	Less than 3.6	$CN_I = \dfrac{CN_{II}}{2.3 - 0.013 CN_{II}}$
II	1.3 to 2.8	3.6 to 5.3	From original CN method
III	More than 2.8	More than 5.3	$CN_{III} = \dfrac{CN_{II}}{0.43 - 0.0057 CN_{II}}$

The following points are important to consider when utilizing the curve number method:

- *Ia*, which consists of interception, initial infiltration, surface depression storage, evapotranspiration and other factors, was generalized as 0.2S based on data from agricultural watersheds (S is the potential maximum retention after runoff begins). This approximation can be especially important in an urban application because the combination of impervious areas with pervious areas can imply a significant initial loss that may not take place. The opposite effect, a greater initial loss, can occur if the impervious areas have surface depressions that store some runoff.

- Runoff from snow-melt or rain on frozen ground cannot be estimated using these procedures.

- The curve number procedure is less accurate when the runoff is less than 0.5 inch. As a check, another procedure should be used to determine the runoff.

- The SCS runoff procedures apply only to direct surface runoff: large sources of subsurface flow or high ground water levels that contribute to the runoff should not be overlooked.

- When the weighted curve number is less than 40, use another procedure to determine runoff..

2.6.3 Hydrodynamic Modeling in Urban Drainage

After the runoff volume is determined, it needs to be transported to its final discharge points. In urban areas the runoff is transported by streets, channels and underground pipes. The movement of the water through the drainage system is regulated by physical processes. These processes are embodied in the well known principles of conservation of momentum and mass. In order to use such physical principles it is necessary to translate them into mathematical equations than can be instantiated by engineers and practitioners. (for further details see for example, Price and Vojinovic, 2011; Vojinovic and Abbott, 2012).

A mathematical model is a representation of the observed reality. The set of equations which describe fluid motion in three dimensions are the Navier-Stokes equations (Eq 27 and 28), namely momentum and mass conservation for an incompressible and homogeneous fluid.

$$\frac{\partial u}{\partial t} + (u \cdot \nabla) u + \frac{1}{\rho} \nabla p - \nu \Delta u = g \qquad (27)$$

$$(\nabla) \cdot (u) = 0 \qquad (28)$$

where:

u is the 3D velocity vector

t is time

p is pressure

ρ is fluid density

ν is viscosity

g is the gravitational acceleration

Analytical solutions of such a set of equations are limited and the few solutions are based on simplifications, which may not be usable in real engineering problems. However, there are numerical methods that can be employed to obtain solutions with sufficient accuracy for practical use. The main drawback is the lack of knowledge about turbulence. When the flow is laminar, low Reynolds number, the flow tends to be deterministic; for large Reynolds numbers, the flows becomes turbulent and stochastic. Turbulent flow mechanisms are still not well described; they can be introduced as random components to the velocities and the equations integrated over the time, as proposed by Osborn Reynolds. This generates what are called the Reynolds stresses which have to be related to the mean velocities. In other words, this approach solves part of the problem but the equations have lost their generalization, requiring then to be complemented with empirical data (Rodi 1993).

The use of the three dimensional Navier-Stoke equations in urban drainage is not common and has been limited to some specialized structures at the expense of high computational requirements (Clemens 1998). Practitioners in the past have been reluctant to take into account time as a variable and prefer more simple models like the Manning equation for steady state uniform flow. It was, and still is, a common practice to compute peak discharge (using the rational formula) and to design the pipe or drainage channel working to full capacity. This approach tends to overestimate the structure sizes under conditions where the flow controls are upstream and to underestimate the sizes when the controls are downstream and a backwater curve is formed. A better approximation than the assumption of steady state uniform flow is the use of a steady state non-uniform flow model. It allows the estimation of backwater effects and gives a better picture of the water levels, but it is still incomplete and lacks the simulation of important features such as the effect of storage in manholes or retention ponds.

Nowadays, the use of one dimensional models solving the full hydrodynamic equations is common in urban drainage. The numerical procedures to solve such equations are mature; they has been used for a long time with an acceptable performance. However some limitations still remain. The use of models based on the two dimensional equations has not been extensively used in urban drainage. This is because practitioners have been focused only on the below-ground system assuming that it should be able to drain the complete area. Now it is accepted that also the above-ground system should be used to drain part of the rainfall excess, and the use of two dimensional models has become of increasing importance. The following sections are aimed at describing the background behind the one and two dimensional flow equations and model instantiation for the urban drainage.

2.6.4   Saint Venant equations (1D)

The usual way to perform flow simulations in urban areas is through one-dimensional models (1D). Computational methods in 1D were originally used to model pipes and conduits in steady and quasi-steady state. Since then, they have evolved into more complex methods that consider the fully dynamical 1D Saint-Venant equations (1871); see Eq 29 and 30. The equations can be derived starting from the 3D Navier-Stoke equations introducing the following assumptions:

- The flow is in one-dimension, the velocity is uniform over the cross section and velocity in y and z (vertical) direction are negligible.

- The streamline curvature is small and vertical accelerations are negligible, hence the pressure is hydrostatic and the water level across the section is horizontal

- The effects of boundary friction and turbulence can be accounted for through resistance laws analogous to those used for steady state flow

- No lateral flow

- The average channel bed slope is small

$$\frac{\partial Q}{\partial t}+\underbrace{\frac{\partial}{\partial x}\left[\beta\frac{Q^2}{A}\right]}_{(b)}+\underbrace{gA\frac{\partial h}{\partial x}}_{(c)}+\underbrace{gA\cdot c_f\frac{QQ}{R_h A^2}}_{(d)}=0 \qquad (29)$$

$$\underset{(a)}{\underbrace{\phantom{\frac{\partial Q}{\partial t}}}}$$

$$\frac{\partial Q}{\partial x}+B(h)\frac{\partial h}{\partial t}=0 \qquad (30)$$

where: Q is the discharge

x is displacement on axis x

t is time

A is cross section area

B is cross section water width

β is Boussinesq's coefficient

g is the gravitational acceleration

h is the water level

C_f is the resistant coefficient

R_h is the hydraulic radio

In Eq. 29 the first term (*a*) is a local acceleration term, the second term (*b*) represents the convective acceleration of the fluid, the third term (*c*) is a gravitational term, and the fourth term (*d*) is the wall friction term. If the full *dynamic wave* model has to be set up, all the terms (*a,b,c and d*) have to be used. Otherwise, simplifications can be introduced if needed. For instance, when the friction is small compared with inertial (*a and b*) and gravitational forces (c), the friction term (*d*) can be neglected in Eq 29 leading to the so called *gravity wave equation*. If the friction term is a large compared with inertial terms the equation is called the *diffusive wave equation* (*c and d*) or the *kinematic wave equation* if only part of term (*c*) is used. The Saint Venant (or shallow water) equations can be written in different ways. They have different forms and different approximations: algorithmic, characteristic, discharge and approximate form.

Friction term

As stated above the solutions of the Navier-Stokes equations, and their simplifications, need to be complemented by empirical equations. One such simplification is made to estimate the losses due to the boundary friction. Most of the forms for friction equation have been developed under normal steady state flow. The friction resistance coefficient (C_f) can be expressed as a function of Darcy friction factor as shown in Eq 31a. The friction factor can be estimated using the well-known Colebrook-White equation 32. There are other alternatives to estimate the resistance coefficient, for instance the Manning equation

31b.

$$C_f = \underbrace{\frac{f}{8g}}_{a} = \underbrace{\frac{n^2}{R_h^{1/3}}}_{b}$$

(31)

and the Colebrooke-White equation:

$$\frac{1}{\sqrt{f}} = -2 \cdot \log\left[\frac{k_e}{12R_h} + \frac{2.51}{R \cdot \sqrt{f}}\right]$$

(32)

where: f Darcy friction factor

n Manning roughness coefficient

k_e Nikuradse's equivalent sand roughness

R Reynolds Number

R_h Hydraulic radio

2.6.5 Numerical Solution of Saint Venant Equations

Analytical solutions of Saint Venant equations are limited to simple cases only. One of the methods that can be used is the Method of Characteristics (MOC). This is a semi-graphic method that can be applied to solve hyperbolic partial differential equations (PDEs). It is based on defining the characteristics along which disturbances propagate. Characteristics are lines in time-space such that along them certain properties remain constant. MOC algorithms for urban drainage network implementation are not straightforward and other methods are normally preferred.

Numerical methods can transform the continuous differential equations into a discrete set of algebraic equations than can be solved algorithmically using the computer. There are four main numerical approaches to solving partial differential equations:

- Finite Differences Methods (FDM): These represent the problem through a series of values at particular points or nodes. Most of the FDMs replace the derivative terms in the Saint Venant equations using truncated Taylor series expansions. The series expansions representing the derivatives can be posed in explicit or implicit form, this last form is preferred for stability reasons. The FDMs are easy to implement but must require a regular discretization.

- Finite Elements Methods (FEM): The basis is to divide the domain into elements, triangular or quadrilaterals, and place in the elements nodes at which the values will be computed. The solution at any position is represented by a series expansion of nodal values within the local vicinity of that position. The nodal contributions are multiplied by an interpolation function.

- Finite Volume Methods (FVM): These divide the domain into a finite number of volumes (cells) but the discretization is done in a integral form of the partial differential equations. The resulting unknown expressions are similar to the FDM methods. The main FVM characteristic is that it keeps an exact conservation of mass within cells so for that reason they have become popular to solve fluid flow problems.

- Spectral Methods: These methods are a variant of FEM with the difference that the interpolation function is not local but global.

Boundary and Initial Conditions

It is stated above that the Saint Venant equations are hyperbolic partial differential equations. Where the Saint Venant equations are applied to a pipe it is important to establish the appropriate boundary conditions for the solution space. This means that initial and boundary conditions are needed in order to have a well posed mathematical problem. The initial conditions state that for an initial time $t=0$ the discharge $Q(t=0,x)$ and water level $h(t=0,x)$ must be known along all of the x direction. In the case of the boundaries the condition will depend on the flow regime.

The regime of the flow is determined using the Froude number (Fr). When the Froude number is less than one, the flow regime is sub-critical and when it is more than one the regime is super-critical. Under sub-critical flow conditions two boundary conditions are need, one upstream and the other downstream. This is because under sub-critical conditions water perturbations can travel down or upstream; this is opposite to super-critical flow in which perturbations can only travel downstream. In the case of super-critical flow again two boundary conditions are needed but this time both have to be at the upstream boundary.

In the case of urban drainage networks many structures are present between pipes introducing discontinuities in the equations. In such cases the boundaries are said to be internal. One extreme but frequent occurrence in urban drainage networks, is the formation of hydraulics jumps: the flow goes from super-critical flow to sub-critical flow, which introduces a discontinuity in the free surface inside the pipe network. In such a case it is necessary to define three boundary conditions: two boundary conditions upstream and one downstream; however, most of the models do not manage such an option. The boundary conditions are usually defined as functions of time; in other words, the values at the boundary must be known from the initial to the final time of the simulation. Common boundary conditions are:

- Discharge hydrograph at a node $Q(t)$ (inlets, catch-pits)

- Water level $h(t)$ in nodes (tidal controlled outlets)

- Discharge-level relations $Q(h)$ (outlets with normal flow, controlled by weirs)

- Internal boundaries controlled by pump systems, weirs, manholes and other structures.

- Closed boundaries $Q(t) = 0$

Finite Differences Discretization

Amongst the many different schemes there are two that are the most wide spread for urban drainage network models. They are: the four-point implicit-difference scheme (Preissmann scheme) and the six point implicit-difference scheme (Abbot-Ionescu scheme). Other schemes are available (see, e.g. Cunge et al., 1980, and Ball, 1985); however, we focus on these two implicit techniques.

Four Points Scheme

The Preissmann or four point scheme is an implicit finite difference scheme that is based on computations on a grid of four points (Figure 2.9). The scheme computes the differences using the set of equations:

$$f(x,t) = \frac{\theta}{2}(f_{j+1}^{n+1} + f_j^{n+1}) + \frac{(1-\theta)}{2}(f_{j+1}^n + f_j^n)$$

$$\frac{\partial f}{\partial x} = \frac{\theta(f_{j+1}^{n+1} - f_j^{n+1}) + (1-\theta)(f_{j+1}^n - f_j^n)}{\Delta x} \tag{33}$$

$$\frac{\partial f}{\partial t} = \frac{f_{j+1}^{n+1} - f_{j+1}^n + f_j^{n+1} - f_j^n}{2\Delta t}$$

In Eq 33 $f(x,t)$ and f represent both the discharge (Q) and the water level (h) which are computed at the same time step for each node. Subscript n is the time step for the current indicating current time and $n+1$ is one time step further; the same applies in *the x* direction which is represented by subscript j. A weight parameter is used (θ) in order to give stability to the discretization. This value ranges between 0.5 and 1.

For single elements (pipes) with boundary conditions given at each end, the coefficient matrix contain elements only in a band along the diagonal. A computational point j is not linked to all other points but only to adjacent points $j-1$ and $j+1$. Due to the diagonal form of the matrix a double sweep algorithm is used for any time step Δt. This saves computer storage and speeds up the computations.

The main features of the implicit Preissmann scheme are as follows:

- It is unconditional stable for values of θ >0.50, the time step Δt can be chosen to match the modeler's requirements.

- The space intervals Δx may be variable, giving the possibility to use non-uniform grids.

- Both, discharge and water level are computed at the same points of the computational grid. It makes it easier to define the stage-discharge boundaries without major modifications.

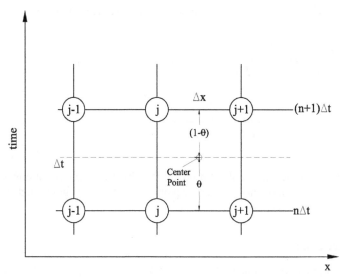

Figure 2.14: Four points FDM scheme discretization according to Preissman

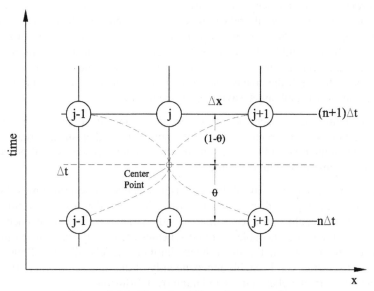

Figure 2.15: Six point Abbott-Ionescu scheme

Six Point Abbot-Ionescu scheme

The most popular six point scheme was introduced by Abbott and Ionescu in 1967 (Figure 2.15). It is defined on an implicit staggered grid, where discharge and levels are computed at alternate nodes. The partial derivatives respect x and t are represented by:

$$\frac{\partial f}{\partial x} = \theta \frac{f_{j+1}^{n+1} - f_{j-1}^{n+1}}{2 \cdot \Delta x} + (1-\theta) \frac{f_{j+1}^{n} - f_{j-1}^{n}}{2 \cdot \Delta x}$$

$$\frac{\partial f}{\partial t} = \frac{f_j^{n+1} - f_j^{n}}{\Delta t}$$

(34)

Just as the Preissmann scheme, substitution of the discretization equations in the Saint Venant equations leads to a linear system of equations with a tri-diagonal matrix that can be solved using the double sweep algorithm.

Stability and Accuracy

The derivatives solved using the numerical schemes are not precise; they are only an approximation to the real analytical solution. The origin of the numerical schemes is an expansion of the Taylor series in which higher (non linear) terms have been truncated. For this reason they do not represent the real Saint Venant equation precisely.

Stability is related to the concepts of consistency and convergence. In numerical methods the size of the spatial or temporal grid can be changed and adjusted by the user or according to computational need. Theoretically, when the grid size tends to zero the numerical differential equation should be the same as the original ones. Consistency in the numerical method means that for a finer grid resolution the methods must be closer to the real solution. In order to check the consistency of a numerical method a test with different grid sizes can be done or by applying a Taylor series expansion.

When a numerical solution scheme of the Saint Venant equations, for a given set of boundary conditions, converges to the same solution as that of the real continuous differential equations, it is said that the scheme shows convergence. A well posed problem must show both consistency and convergence.

Numerical schemes can expose instabilities. This means that the solutions must be bounded in an range between steps in displacement or time; it means that a small perturbation in a boundary condition could produce a increasingly large perturbation at each time step which will make the model unstable. Explicit schemes are usually stable when the Courant number is less than one. On the other hand, implicit models, like the Preissmann and Abbott-Ionescu schemes can be stable independently of the time step. Other errors that can affect the accuracy are the phase errors and numerical diffusion.

$$u \frac{\Delta t}{\Delta x} \leq Cr \tag{35}$$

where:

Cr is the Courant number

u is velocity

Δt is the time step

Δx is the distance interval

An extensive description of the accuracy and stability problems encountered in various finite-difference schemes can be found in many publications e.g. Miller and Yevjewich, (1975), Cunge and Abbott (1986), and Abbot et al. (1979).

In the case of sewers, other sources of loss of accuracy or instability are due to the numerical algorithm not being able to manage all possible cases of flow and sewers in very complex systems. Five types of instability according to the type of flow in sewer systems can be enumerated (Yen and Pansic 1980) :

- Transitions from open channel flow to close dconduit flow and vice-verse

- The transition between super-critical and sub-critical flow in open channel flow

- Roll-waves in super-critical flow

- Near dry bed flow, and

- Surcharge flow

When the pipe is close to working at full capacity the fluid starts to behave chaotically. This is mainly generated by air entrainment which promotes a two-phase flow when the water level is near of the pipe crown. This can start fluctuations between open and pressurized flow. This is viewed as a boundary condition problem in the numerical method because there is no unique level-discharge relationship.

Super-critical flow introduces changes that cannot be managed by a unique algorithm. Transition from super-critical to sub-critical flow produces a hydraulic jump inside the pipe and therefore a discontinuity in the equations. FDMs applied to the discretization of the full Saint Venant equations are not able to manage such discontinuities without a considerable amount of additional coding. Models based in FDMs can deal with sub-critical or super-critical flows, changing the boundary conditions (two upstream for super-critical and one upstream and one downstream for sub-critical) and using the double sweep or single sweep algorithms respectively. However, they cannot work under transitional flow. Some implicit schemes have been proposed to solve this problem but they are not adopted in commercial software see (Kutija 1993), (E. A. Meselhe and Holly Jr. 1997), (Lyn and Altinakar 2002), and (Djordjevic et al. 2004) for a detailed description of transitional flow. Some approaches consist in combining algorithms or in simplifying the full Saint Venant equations using the non-inertial equation (Diffusive Wave equation).

The dry bed problem arises due to a singularity in the numerical equations at low flows. This singularity is related to the boundary and initial conditions which produce discontinuities in the fluid (see Cunge et al 1980). Most of the commercial software products based on FDMs avoid this singularity in the continuity by adding a small amount of fluid at the upstream end of the pipe and then removed at downstream end. This can lead to an imbalance in volume between the rainfall input and the outflow volumes. However, usually this extra amount does not affect the results. In the case of a combined sewer system, the amount of dry weather flow is usually enough to maintain an amount of fluid in most pipes. Another alternative is the use of the inverse Preissmann slot (Ehab A. Meselhe and Holly 1993) as described below. Both methodologies are suitable to use with finite-difference schemes.

One of the most common problems in sewer systems that can affect the accuracy of the numerical scheme is pipe surcharge. When a pipe is surcharged a wave travels from downstream to upstream forming a shock wave. This wave forms a discontinuity and marks a line between free surface and pressurized flow. When the flow changes to pressurized flow the governing equations change slightly; pressure and friction becomes more important than gravitational effects. In order to model transient flow in pressurized pipes the equations in Eq. 36 are used; the cross section area becomes a constant and the energy terms are reformatted:

$$\frac{1}{g}\frac{\partial V}{\partial t}+\frac{\partial}{\partial x}\left(\frac{\beta V^2}{g}+\frac{P_a}{\gamma}\right)=-S_f$$

(36)

$$Q=A_f \cdot V$$

where:

V	is the mean velocity in the cross section
A_f	is area of the full cross section
P_a	is the pressure
γ	is the specific weight of the fluid
β	is a kinetic energy coefficient
g	is the acceleration of gravity
S_f	is the energy slope

However, the change to the original equation leads to a modification of the numerical scheme and the corresponding algorithm. In order to avoid changing the algorithm it is common practice to introduce a hypothetical slot along the crown of the conduit (Figure 2.16). This concept was introduced by Preissman in 1961 (Cunge, 1980). The slot should be narrow in order to have sizeable velocities for the dynamic waves and to minimize continuity problems. However it cannot be too small to minimize numerical problems due to the high velocity of disturbances in the slot:

$$B = g \frac{A}{a^2} \tag{37}$$

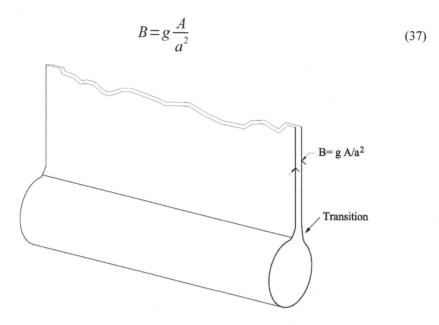

Figure 2.16: Circular pipe with hypothetical slot

Theoretically the recommended width (*B*) of the slot is given by Eq 37 where *A* is the area of the cross section (without excess area due to the slot) and *a* is the wave celerity of a disturbance in the slot. The value of *a* is around the speed of the sound in water combined with other effects like the elasticity of the pipe and air content of the fluid; it is in the order of 1000 m/s. When pipe diameters are small, Eq 37 gives very small values introducing numerical problems (Cunge, 1980). For such cases it is recommended to use values of between 0.1% to 5% of the pipe diameter and not less than 1 cm. This may introduce accuracy problems.

The use of the Preissmann slot can introduce numerical problems and inaccuracies, as stated above. However, it has advantages that make it very useful and its use is promoted in most of the sewer system commercial software products. For instance the solution algorithm remains the same, it is simple to implement, and it can give good approximate solutions for transition between free surface and pressurized flow.

Sewer Structures Modeling

It is common to find structures, other than pipes and channels, in urban drainage networks; for instance, manholes, weirs, storage tanks and pumps. In order to take into account such kind of structures in the numerical model they must be described in term of mass balance, momentum or energy equations. The effects on the flow conditions of such structures occur in small space distances compared with the flow in channels and pipes; for such a reason these structures are modeled as a discrete points in the system (M.B. Abbott and Cunge 1986). It is expected that these structures (discrete points) link with the pipes (continuous links) in such a way that the algorithms can provide rapid solutions without iteration. These feature also impose internal boundary conditions in the system.

The pipe junctions or manholes are the most common discrete feature in a sewer system. They are evaluated using the continuity equation to account for storage inside the manhole. The energy losses that they produce are expressed as functions of the kinetic energy and a loss coefficient (Equation 38).

$$h_L = \frac{K_L \cdot V^2}{2g} \tag{38}$$

where: V is the velocity

                  K_L is loos coefficient

                  h_L is the total energy loss

                  g is the gravitational acceleration

Energy losses in manholes depend of a set of variables that are functions of the flow regime in the approaching pipes (sub-critical or super critical), the diameter and discharge of the pipes, the incidence angle of inflows and outflows, and the volume of the manhole among others. The junctions can be classified as follows (Mays 2001):

- According to geometry: One, two, three, four or more way junctions. They can be merged (most of the pipes merge into one pipe), or can be divided, meaning that one pipe flows into two or more pipes.

- According to the flows in the joining pipes: a free surface flow junction, a partially surcharged junction or a completely surcharged junction.

- According the storage capacity of the junction: point junctions (zero storage) and storage junction.

As can be seen there exist many possibilities that can be present in the sewers, and each is a particular case. Head losses for manholes can be found in the literature (see (Marsalek 1984), (Butler and Davies 2004), (Mays 2001), (Shinji and Tetsuya 2005) . Special cases can be treated separately depending on the flow conditions; for instance, surcharged manholes have different head loss coefficients compared with free surface manholes (Zhao et al. 2008). Air effects in manholes also have a big impact on the losses estimation and have to be taken into account; large peak pressures and severe pressure oscillations can be produced inside the system during storms (Zhou et al. 2004).

Other structures may exist and they also are modeled as discrete points. Weirs are represented through a stage-discharge curve as an internal boundary condition; they can convey water from one node to another. A pump is represented by a characteristic discharge curve; it extracts water from a manhole or storage tank and conveys the output to another node. An example of an internal boundary condition is shown in Figure 2.17.

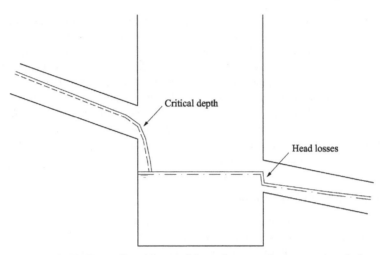

Figure 2.17: Example of internal boundary condition in a manhole

2.6.6 Model Instantiation

The main idea behind modeling is to have a reliable and safe model which the user can trust. Simulation software products by themselves are not models; they are no more than a set of instructions that depend on certain parameters. In order to have a reliable model it is necessary to instantiate it from the software template. Once the software has been selected and the objective of the project has been set, the model has to be built. In order to build the model it is necessary to fill-in the properties of the software template. Some of these properties are built in already in the software and others have to be set by the user.

A model is composed of procedures and data. In order to instantiate the model the following steps are required:

- Collect the required data

- Build and calibrate the model

- Verify the model

Data collection

The model relies on the quality of the data that is used for building. Urban drainage models require primary categories of data: geometrical or structural, parameters and performance data. Geometrical information is obtained from surveys and describes the geometry of the model structures. The values of model parameters are derived from data such as land use for infiltration or CCTV for the roughness of pipes. Performance data is aimed to test and validate how good the model is. It includes rainfall, flows overland and in channels, water quality and sediments.

Also, the data is sub-divided into above-ground data and below-ground data. Above-ground data is aimed at building the hydrological rainfall-runoff model; however, as will be seen in the following section, it can also be used to build the hydrodynamic model when dual modeling is required. The above-ground data is related to the catchment delineation, overland flow, land use, overland flow paths, street geometry, population, among others parameters. In the case of below-ground data what is needed are pipe diameters, lengths, slopes, terrain levels, diameters and head loss coefficients in manholes, structures such as weirs, pump systems, etc. Below-ground data acquisition is more uncertain than for above-ground. Usually some information has not been surveyed or its existence is ignored and its absence can affect the model performance.

Model Calibration and Verification

Once the data has been collected and checked, the model has to be built. The user must decide what data to put into the model, select the values for the parameters in use based on the available information, and what simplifications are allowed and their influence on the model results. It is impossible to build model that reproduces exactly the real world; however the user can set up a model which fulfills the aims of the project.

The next step is model calibration. Calibration is the process in which the values of the model parameters are adjusted to closely reproduce the observed data; this can be a very time consuming task. Geometrical or structural data usually do not need to be adjusted, but only those parameters that are not directly measured or parameters with high uncertainty their estimation which makes them suitable for adjusting in the calibration process. The part of the model more appropriate for calibration is the rainfall-runoff model. Rainfall is a stochastic variable that changes in time and space; the output of such a process, the runoff, is a stochastic variable as well (M.B. Abbott and Cunge 1986); however they are treated as deterministic input and output in urban drainage modeling. Usually, the coefficient in the rational equation, the infiltration rate, the impervious area connected to the system, or the curve number (depending on the model) are the first parameters to change in the calibration of urban models. Also the shape of the catchment areas can be moved if is needed. The hydrodynamic model tends to be more deterministic than hydrological models; for such a reason the parameters to calibrate are reduced to roughness and local energy losses in the structures. Other calibration parameters can exist depending on the instantiated model; for instance, when there is a pump system the characteristic curve may require adjustment due to aging of the pump.

Calibration requires us to compare observation and simulated data. In order to compare these an error measurement indicator has to be selected. The error indicator will depend on the objective of the model. If peak accuracy and arrival time is important an error indicator based in these variables must be used. There are many error indicators that can be adopted (see Mays 2001).

The model has to be verified in order to warrant generalization of the model. The verification consists of a comparison of the model, using an error indicator, with new observed data that has not been used for calibration. It will assure that our model is a representation of the real phenomena. At this stage no parameter adjustment has to be done; if our model does not perform correctly the user has to analyze the possible causes of such poor performance and return to the data collection or software selection stages. Data for urban drainage model calibration is scarce and their collection is costly to obtain, so the questions that arise are: why not use this data also for calibration after the model verification is acceptable?. This may produce a better model than that originally calibrated.

2.6.7 Below and Above ground Modeling

Models can deal with the flow in pipes (underground flow) and channels (above-ground flow), but they are weak when surcharging in underground conduits causes flooding of the streets. In such cases the 1D model normally manages the flooded area as a temporary storage that returns the water volume to the sewerage system later on. The limitations of 1D models for UDM have been discussed by Mark (2004). He points out that a new approach is needed with 1D modeling for underground flows and 2D modeling for above-ground flows in order to get a better reproduction of the real world system. A study of data requirements for this situation is also discussed. Furthermore, a study by Vojinovic and Tutulic (2009) suggested that since the 1D above-ground approach does not account for 2D flow processes, it places more responsibility on its users as it requires extra effort to define the terrain geometry and overland flow paths.

Surcharge in sewerage systems has been addressed by several researchers. Djordjević (1999) introduces the concept of dual modeling, using the full Saint Venant equations in the sewer system and simplified methods for the above-ground flows when the capacity of the underground network is exceeded.

Hsu (2000) describes the problem of flooding due to sewer surcharge. An urban inundation model was developed combining a 2D diffusive overland-flow model with the SWMM model (1D), including pumping stations. The model was tested in areas of downtown of Taipei City using both a real recorded event and a 100 years rainfall. The model helped to identify potential inundation areas.

Schmitt (2004) developed an integral tool for planning and management based in cost evaluation. The procedure uses the ResUrSim model, which considers flooding due to the surcharge of the sewerage system. The tool is applied to a small area in the city of Kaiserslautern. Since no flooding measurements were recorded in the period of study synthetic rainfalls were used. Particular attention was given to the detailed surface and sewer flow interaction. Important developments were made in hydraulics surface simulations, the coupling of surface and sewer flow and the use of GIS in data collection.

The ReSurSim model uses the full Saint Venant equations for the underground system (1D). Surface modeling is carried out by solving the simplified shallow water equations neglecting the inertia terms in the momentum equation. An explicit difference scheme is used.

Mignot (2005) applied a 2D model based on the shallow water equations to simulate an urban flood. The model was tested using a physical model that represents the flooding of the center of Kyoto due to overflow from the Kamo River. The numerical 2D model represents the depth and flow rates fairly well but there are discrepancies at crossroads.

It can be argued that modelling in urban drainage is going towards the introduction of 2D – 3D aspects of the flooding. However, this approach is still ongoing. One important issue is the extent to which the real 2D behavior of flooding can be modeled. Commercial and research developments are moving in this direction, SOBEK (from Delft Hydraulics) has been extended to include a 2D connection with 1D support. Similarly Danish Hydraulic Institute (DHI Water and Environment) has also provided an experimental connection between MOUSE and MIKE-21.

Open source code is also available; FESWMS use finite elements to couple 1D-2D schemes. However such software in UDM cases has not been extensively applied. More practical evaluation of the influence of small channels in urban inundation areas is needed. In the following section a number of packages for urban drainage modeling are enumerated.

2.6.8 Existing urban drainage modeling systems

- SWMM (1D): The Storm Water Management Model (SWMM) is a comprehensive water quantity and quality simulation (1D) model developed primarily for urban areas. Single-event and continuous simulation can be performed for almost all components of the rainfall, runoff, and quality cycles for a watershed. The Extran Block of WMM also includes complete dynamic flow routing for hydraulic simulation of dynamic backwater conditions, looped drainage networks, surcharging, etc. Public domain source code is available.

- SOBEK URBAN (1D-2D Coupled): Sobek is a 1D-2D integrated model, it solves the Saint Venant flow equations and the fully 2D shallow water equations in an implicit way using finite differences. It works for subcritical flow and it is capable to deal with wetting and drying areas.

- MOUSE (1D): The computation is based on an implicit, finite difference numerical solution of the basic 1D, free surface flow equations (Saint Venant). Subcritical flows can be simulated. Free surface and pressurized flows are computed within the same algorithm using Preissman scheme. The MOUSE Surface Runoff Module includes three types of surface runoff computation: Time-Area Model, Kinematic Wave Model and Linear Reservoir Model. It also has an additional option to model slow response component due to groundwater infiltration, namely Rainfall Dependent Inflow Infiltration (former NAM and present DRII model). Mouse also has integrated water quality and sediment transport models. The models are integrated using a GUI-GIS application.

- InfoWorks CS (HydroWorks) (1D): This package consists of a single environment that integrates hydraulic modeling with comprehensive data management and links to GIS systems. The model is 1D using the Priessmann 4-point scheme to approximate the Saint Venant equations and can analyze both steep and flat sewers or open channels. It also deals with subcritical flow and wet-dry conditions.

- ResUrSim (1D-2D Coupled): This is a dual drainage model which uses a 1D fully hydrodynamics module (HamokaRis) for underground simulations and a 2D module (RisoSurf) that solves the shallow water equations neglecting the inertia terms. Both models are coupled through the manholes and catch pits. The model was developed by ITWM (Dr. K Nieschulz and Dr. N. Ettrich).

2.6.9 Generic 2D Modeling systems

- FESWMS: is a finite element hydrodynamic modeling code that supports subcritical flow analysis, including area wetting and drying. The FESWMS model allows users to include weirs, culverts, drop inlets, and bridge piers in a standard 2D finite element model. Source code is available in FORTRAN 77 language.

- RMA2: is a sub-critical, depth-averaged model for both steady state and transient hydraulic modeling. General applications require an unstructured finite element network of quadratic elements (6 node triangles and 8 node quadrilaterals). Sections of one-dimensional entities, which simulate trapezoidal channels can be incorporated into the two dimensional mesh.

- HIVEL-2D: supports sub-critical flow analysis. It is used to analyze flow fields, which have shocks such as hydraulic jumps and oblique standing waves. SMS supports both pre and post-processing for HIVEL-2D.

- MIKE 21: is an engineering software package containing a comprehensive modeling system for 2D free-surface flow. MIKE 21 is applicable to the simulation of hydraulic and related phenomena in lakes, estuaries, bays, coastal areas and seas where stratification can be neglected.

2.7 Summary

A review in the state-of-the-art in urban drainage rehabilitation has been done. During a rehabilitation process several aspects have to be addressed. Issues as determination of performance indicators for hydraulic, structural and environment assessment have to be done, data availability, identification of critical pipes and channels are of major importance in any rehabilitation plan. Hydrological and hydrodynamic modeling plays a key role during for the hydraulic, structural and environmental assessment. Dual modeling of above and below-ground system is preferred in order to evaluate surcharge consequences for the different kinds of the needed assessments. There exist very well mature models for 1D and 2D modeling, however, the interaction between them is still matter of research to become available for practitioners. Sustainable approaches, oriented to control runoff volumes since the beginning of the rainfall are preferable than methodologies based on conveyance. These sustainable approaches are also oriented to keep environment, social and economical values in balance. Different models for hydrodynamic modeling were reviewed in order to select the most suitable for this research, indicating advantages and disadvantages.

3 A Multi-tier Framework for Urban Drainage Rehabilitation

This chapter introduces a proposed multi-criteria framework for further development of a decision support system (DSS) for urban drainage rehabilitation. Its theoretical basis and the connection and interrelation between its components are presented. Also, the prototype of the multi-tier framework optimization framework is presented.

3.1 Introduction

Despite the considerable knowledge generated in the area of urban drainage planning and management, there are still many challenges concerning the design and maintenance of urban drainage systems (UDS). The different phases of flow in a UDS are traditionally managed separately, including flow in sewers, overland flood flows, treatment works, groundwater, and receiving waters. Understanding the relevant physical and chemical processes and using the appropriate model for each phase is of paramount importance. Model integration with multi-objective optimization techniques opens up a range of options for a holistic approach to UDS including full utilization of the existing system's capacity prior to undertaking improvement works, the reduction of pollution affecting local receiving waters as well as main watercourses and groundwater, the reduction of infiltration/inflow to the network and its adverse impact on waste-water treatment, the minimization of damages due to surcharge affecting properties and the environment, limiting the risk of structural failure and damage to other sub-surface infrastructure, the prevention of the pollution of water supply systems due to the infiltration of waste-water, and the minimization of expenditure.

Furthermore, the rehabilitation process should allow for an input by stakeholders. However, when decisions have to be taken, the interests of a wide variety of stakeholders make the interpretation of drainage information very difficult. The involvement of many interests usually leads to conflicting objectives. In such cases, managers and system planners need effective methodologies and tools that provide a holistic view in order to come up with an optimal trade-off between different interests. Inevitably, this requires the use of multi-objective approaches and tools to deal with the conflicts that arise.

3.2 The Fifth Generation of Modeling System

Abbott (1991) describes the historical evolution of computational hydraulic into hydroinformatics in four well defined generations. The first generation was born with the first computers in 1950s. The computer was used to solve easy analytical equations in the same way done by humans. The second generation introduced the first numerical approximations to partial differential equations in the 1960s. Usually these models were particular to specific cases; during this period some modeling groups started to grow and began work on the improvement of the algorithms. These groups evolved into computational hydraulics centers and started the development of more robust software that could be instantiated in any region for similar application types. These computational hydraulic centers were born in the 1970s and their goal was commercial and gave birth to the third generation of models . In the 1980s there was a technological revolution with the maturity of the PC; the prices were affordable to many people and and hydraulics software started to proliferate. The fourth generation was characterized by the inclusion and integration of other informatics tools such as data bases, sophisticated visualization techniques, geographical information systems (GIS) with the hydraulic software. Separately, other techniques, distinct from numerical methods, has been developed and used in the area of hydraulics during this fourth generation; these techniques are known as Artificial Intelligence (AI).They include Artificial Neural Networks (ANN), Fuzzy Logic (FL), Genetic Programming (GP), Genetic Algorithms (GA), Chaos Theory and Model Trees among others. The integration of both, numerical and AI, techniques and the incorporation of stakeholders in the decision making process is what Abbott foresaw as the fifth generation of modeling.

3.2.1   Towards the Fifth Generation of Modeling for Urban Drainage

Introduced almost 20 years ago, the fifth generation of modeling systems is beginning to be seen today; however, the use of rule based expert systems to guide the modeler may still be some way off. For a long time the need has been identified for expert system and computation tools for decision making in the area of urban drainage and sewerage; see Barraud et al. (1999), Hahn et al. (2002). Most of the tools have been limited to a set of if-then rules in order to guide and help the user in the decision making process, and only recently have more sophisticated tools been developed (Vojinovic et al. 2005), (Saegrov 2006), (Martin et al. 2007). Ana and Bauwens (2007), divide the decision-support tools for sewer network asset management into three main groups.

The first group consists of tools that mainly deal with performance modeling. These tools are based on data collection and analysis of such a data. They also do performance modeling using models of the deterioration, forecasts, demands or simple impact assessments. Some models in this category are: the Hasegawa et al. (1999) (M1) which ranks the pipes based on four variables: decrease in flow capacity, road collapse possibilities, sewer overflow and increase in treatment costs; the Bengassem and Benis model (2000) (M2), which is based on a fuzzy expert system, and takes into account the structural and hydraulic performance, and the vulnerability of the site using a score from 0 to 100; the Baik et al. model (2006) (M3), which predicts the future condition of the sewer using Markov-chains probabilities. In order to estimate the probabilistic functions, inspection of the network is required.

The second group of models include those that use a measure of performance but also incorporate a decision analysis tool. They follows three steps from data collection, to performance modeling and decision analysis. APOGEE in France (M4), AQUA-WertMin in Germany (M5), Edmonton models (M6) and PRISM (M7) in (Canada), KureCAD in Finland (M8) are models in these category. The main characteristic of these models is that they include databases, GIS support, and models to analyze scenarios or alternatives.

The third group of modeling tools has the same characteristic as the second group, but additionally they include reporting, monitoring and management information tools. They also follow the complete infrastructure management stages as presented in Chapter 2. In this category, CARE-S in France (M9) (Saegrov 2006) and HYDROPLAN in Belgium (M10) are good examples (Vojinovic et al. 2005).

Table 8: Relevant data in sewer asset management tools

Tools	Pipe material	Pipe age	Pipe length	Pipe Diameter	Pipe thickness	Pipe Shape	Depth of pipe	Type of pipe	Pipe roughness	Pipe slope	Pipe location	Condition data	Defects History	Flow data	Leakages rates	Soil data	Land use	Traffic road	Tree location	Groundwater data	Population	Geo-technical data	Type of joints	Economic data	Rehabilitation costs
M1	X	X	X	X			X	X		X		X		X	X	X		X		X			X		
M2		X	X				X			X	X	X	X	X		X						X			
M3	X	X	X	X			X	X		X		X		X	X	X		X		X					
M4	X	X	X	X		X	X	X	X	X	X	X	X		X	X	X	X		X		X			X
M5	X	X	X	X		X		X		X	X	X	X		X	X		X					X	X	X
M6	X	X	X	X		X						X				X						X			X
M7	X	X	X	X			X	X				X													X
M8	X	X	X	X							X	X	X	X		X	X	X	X				X	X	X
M9	X	X	X	X	X	X	X	X	X	X	X	X	X	X	X	X	X	X	X	X	X	X	X	X	X
M10	X	X	X	X		X	X	X			X	X	X		X	X	X	X	X		X				X

3.2.2 A multi-criteria approach to optimization

As can be seen the available tools differ in their capabilities and functionalities. However, in terms of complexity most of them are very complex, rigid and require a high quantity of expensive data (Table 8). This situation limits their use on large cities (Stone et al. 2002).Also these complex tools may require several people to implement and run the system. Ana and Bauwens (2007) remark on the need for *"light-versions"* of decision support tools that can easily be applied at any utility level, requiring only available and obtainable data.

Such a simplified approach becomes more important in developing countries, where data is scarce and its collection is costly due to environmental and social conditions. These tools should allow decision makers to take decisions using the Pareto principle, but they should also save scarce resources that can be used in the rehabilitation of the sewer networks. The tools have to be flexible, allowing the introduction of simple or complex procedures for sewer rehabilitation evaluation. If the data is available a complex rehabilitation procedure can be used; otherwise, they can be substituted by more simple procedures. This means that the tool should allow scalability. In the following sections an approach for using such a tool is introduced.

3.3   A Cascade of Optimizers

The rehabilitation tool can be seen as a cascade of optimizers (Figure 3.1) that can be implemented in three stages. In the first place it is necessary to identify spatially where rehabilitation is required. For this reason at first optimization is performed based on the current sewer network diagnostic; this can be seen as a reduction of the domain. Secondly, for rehabilitation a set of alternatives has to be generated and evaluated for the selected area ; it consists of the optimization of the alternatives considering the multi-criteria nature of the sewer rehabilitation process. Finally, the Pareto set obtained in the previous step, containing several optimal alternatives, has to be filtered by stakeholders and decision-makers based on their preferences leading to a third stage optimization.

3.3.1   Domain Reduction

Domain reduction is carried out through a diagnosis of the current system. In order to do this diagnostic a series of indicators for the structural, hydraulic and environmental evaluations have to be selected, as shown in Chapter 2. The selection of such performance indicators has to be in agreement with procedure used. In our case, due to the fact that the methodology and the respective optimization tool will be applied to developing countries, a set of simple indicators was selected. However, the procedure is also suitable for more complex tools and applications as stated in the section above.

The main goal here is to help practitioners to identify a subset of pipes suitable for change. This is done through a performance analysis that is mainly carried out in two stages. The first stage is semi-automatic and addresses four aspects: hydraulics and environmental performance, structural conditions, and the identification of strategic pipes.

The hydraulic performance is determined through the hydrodynamic model using the current conditions. The relation between the pipe capacity and maximum flow in a pipe (Q/Qf) is used to find those pipes that do not perform well. The ratio between the current discharge and the pipe capacity is an indirect measure of the surcharge in the pipe, which is the main cause of backwater effects in the sewer network. Another possible relation is the use of the gradient of energy and terrain slope (S/S_0). The discharge ratio was selected due to its common use among engineers and it is an accepted parameter in many storm-water standards. However, due to the modularity of the methodology other hydraulic indicators can be used such as the one described in Section 2.3.2.

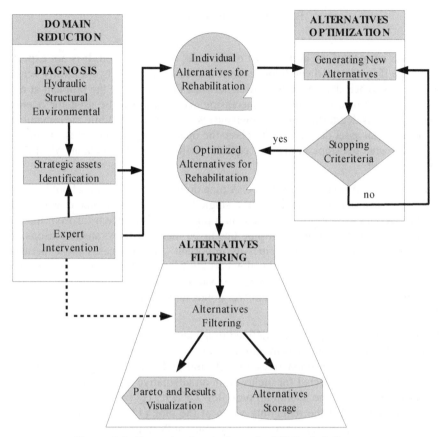

Figure 3.1: Cascade of optimizers for UDS rehabilitation

Structural condition can be assessed using the age of the pipe, visual inspections, the number of repairs per year or expert judgment. The pipes are classified depending of their conditions in classes from 1 to *n* using expert knowledge, field visits and CCTV when it is available. This classification represents not only the aging of the asset, but also its current condition. Assets can be degraded not only by time but also due to use or aggressive environmental conditions. Using this classification the remaining fraction of the asset value can be calculated using the equation:

$$V = S^{k/n} \tag{39}$$

where: V is remaining value of the asset as fraction of initial value

S is recovery value of the asset in fraction

k is the current class of the pipe between 1 and n

n Is the number of classes

Eq. (39) gives an idea of the current cost of the asset if it is replaced. It is not expected to replace a pipe if it is in good condition, even if it does not perform hydraulically very well. There may exist other changes to the sewer network that can solve the problem at a lower cost. Similar equations, but with cost increasing with class, can be used for operational cost; when a pipe is degraded, more maintenance will be required. While the remaining cost of the change in an asset decreases with class, the operational cost increases; if both values are added a change of slope in the new equation will indicate the moment when the asset should be replaced.

Strategic assets are of primary importance. A pipe draining a hospital or school has a higher level of importance than the one draining a single basement. Also a pipe downstream and close to the output has more importance than a pipe located upstream far from the output. There are many situation in which a pipe has to be included for strategic reasons. There could exist many reasons to rank pipes as important; for such cases the experience of practitioners is again required. Additional expert knowledge is required on the location and selection of other alternatives than of pipe renewal. Experts have to locate sites suitable for storage tanks, or the location of dry and wet ponds and other rehabilitation technique that could be used.

For the second stage a set of rules is defined to achieve a suitable and optimal selection of pipes and rehabilitation techniques to introduce in the process of generating alternatives. In order to implement the selection rules suitable threshold values can be introduced. For instance, a pipe will be selected for inclusion in an alternative scenario if its ratio between the current discharge and the maximum pipe flow capacity (Q/Qf) is greater than 1.5. Also a pipe may be selected if the remaining value of the pipe (V) is less than 0.5 times the reposition cost of the asset (C); where C is a function of the pipe diameter.

Also, fuzzy logic rules could be used to combine hydraulic, structural and strategic importance conditions in order to select these pipes. Before starting the optimization, the practitioner should be able to incorporate (or eliminate) the pipes he thinks should be modified. At this stage a GIS tool can be used to analyze the spatial variability of the selected pipes. It is more efficient to modify sets of pipes belonging to the same area than single pipes separated and scattered, avoiding disruptions in several parts of the city. Figure 3.2 shows a GUI for the pipe selection to be optimized, allowing for the inclusion of expert knowledge.

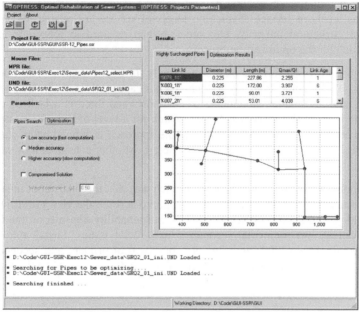

Figure 3.2: Domain reduction with the OPTRESS tool

3.3.2 Optimizing Alternatives

Once a set of rehabilitation measures are identified and selected using the diagnostic procedure, they have to be optimized. The alternatives can include pipe renewal, evaluation, location and dimension of wet and dry ponds, location and evaluation of diversions, location and evaluation of overflows. These alternatives require to be evaluated in order select the best solutions on which decision-makers and stakeholders can negotiate. This optimization process it is not obvious and becomes a titanic task when the sewer network system is complex; the number of total combinations can be estimated as n to the power p (n^p); where n is the number of pipe diameters available in a catalog and p is the number of pipes to change . For instance, in a medium size network it is common to have 200 to 500 pipes. If after a selection process 10 pipes are selected to be rehabilitated using pipe renewal, in a catalog of 8 possible commercial diameters it gives a total of 1,073,741,824 alternatives. If additionally to this combination, two storage tank evaluations and 2 overflow/diversions (and the dimensioning of such rehabilitation alternatives) are added, the number of possible combinations will increase. For these reasons a suitable and realistic optimization approach is required.

Levels of service

In order to assess effectively the performance of an urban drainage system it is important to have a clear definition of its levels of service. The concept of "*level of service*" in a UDS usually relates to risk. However, its definition can be interpreted from a wider point of view. Any degree of satisfaction or agreement about the service provided by a UDS can be measured in quantitative or qualitative terms. These degrees of satisfaction can then be interpreted as levels of service. Typically, levels of service are related to: public health, flooding, risk, structural integrity, etc. and they can be used to formulate an objective function needed for the optimization process.

There are two kinds of performance criteria, namely fixed and variable. The values used for fixed criteria usually come from experience. Such an approach is used when a variation in performance implies smaller increments of costs or when there is a lack of information. Variable criteria of performance involve the search for a level of service that optimizes the balance between benefits and costs. This variable approach is applied when changes in the level of service increase costs, and also when there exist other considerations such as social or political issues.

The rehabilitation of an UDS must be done in such way that the final set of measures forms a part of a sustainable system. This means taking into account the views of different parties and stakeholders and their conflicting interests. Sustainability depends on the interrelation between technical, environmental, economic and social issues. Sustainable criteria are presented by Ellis et al (2004) and they are composed of primary and secondary criteria; see Chapter 2. These criteria are transformed into indicators or objectives that can be optimized.

Multi-tier approach

The multi-objective optimization process discussed here links several layers (tiers) of levels of service, as shown in Figure 3.3. This approach allows a trade-off zone to be set where users and decision makers can negotiate inside a common area of acceptance. The levels of service are assumed to be spatially distributed over the entire drainage area. For instance, flooding damage, risk, aesthetics are spatially distributed.

A multi-objective optimizer can be used to generate various alternatives which would then be evaluated within a suitable hydraulic (1D or 2D) model. The respective levels of service, rehabilitation costs, flood damage or risk, etc, can be computed from the model results for each cell and for each selected level of service. The use of a GIS system supported by its map algebra functionalities can then be applied to integrate the levels of service over the entire urban catchment area. Such a process needs to be repeated until the optimal set of solutions (or a Pareto set) is identified.

In this way, decision makers and stakeholders are able to derive a suitable set of solutions that is essentially optimal and allows for safe negotiations. The main advantage of this approach is that it allows different units to be used for the levels of services and it avoids the need for monetary expressions.

The definition of the objectives or level of services to optimize can be of different kinds. The tool can be adjusted to simulate any objective through the writing of the respective module which is called by the control tool. Chapter 5 presents a detailed formulation of the selected objectives for this study and a proof of concept example.

Multi-objective optimization and its software support

For multi-objective optimization two linked software tools (OPTRESS and NSGAX) have been developed. After a review of optimization methods, it was concluded that the most suitable technique to use was multi-objective genetic (evolutionary) algorithms. They have been applied successfully in numerous studies related to water supply networks. Also there is a very active research community in this field including the development of multi-objective algorithms.

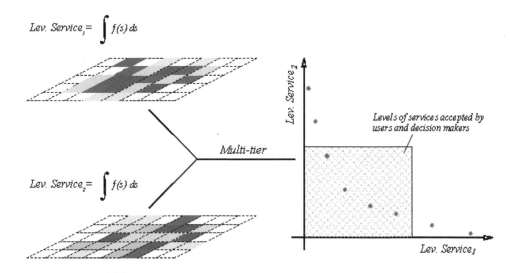

Figure 3.3: Multi-tier multi-objective approach

Two main multi-objective evolutionary genetic algorithms (MOEA) were tested: No-dominated Sorting Genetic Algorithm II (NSGA-II) and Epsilon Multi-Objective Evolutionary Genetic Algorithm (ε-MOEA). Finally, the No-dominated Sorting Genetic Algorithm II was selected due to its robustness, and convergence ability.

Originally, NSGA-II was written for the Linux operating system, so it was ported as a dynamically linked library (DLL in MS Windows and a SO in Linux) by the author to be used in Microsoft Windows, and in Linux as well. Around these DLLs the NSGAX tool was developed (Figure 3.5).

Yet another tool, OPTRESS, tool implements several functions making it easier to select the pipes directly from the hydraulc modelling software database, to launch NSGAX optimizer and to view the results (Figure 3.6). Both tools were developed using the Delphi RAD Environment and Object Pascal programming language.

In order to communicate the modules that evaluate the objective functions and the hydrodynamic model with the optimizer a communication interface layer was developed in Pascal Delphi called Iface (Figure 3.4) . This layer is in charge of translating the NSGA-II population into a set of alternatives to be evaluated. It reads the diameters, storage tanks and overflow/diversions generated by NSGA, and builds and writes a new input file for the hydraulic model. Then, it has to run the hydraulic model and locate the hydraulic results available to be used by the other modules. After that, the interface layer makes a call to the selected modules to compute the objective functions and return the results to NSGA-II. NSGA was originally developed to be run a fixed number of times and to increase its flexibility, we realized the possibility of setting stopping criteria. There exist several Pareto performance indicators, five of them were evaluated in order to select one or a combination of them as a stopping criterion.

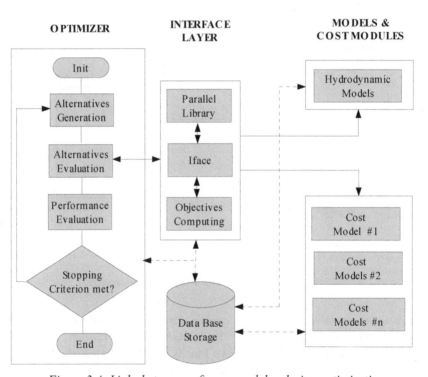

Figure 3.4: Links between software modules during optimization

One of the very important problems to face was the time consumed in the optimization. As it was stated above, a lot of runs has to be done in order to achieve practical results. If the hydrodynamic model takes 30-60 seconds to make a single run it means that for 10000-12000 iterations, which is an average number of runs for optimizing a small size network, it will take around 6-7 days for the whole optimization process. This is one reason that optimization using hydrodynamic models is not wide-spread.

To overcome this problem three approaches were tested. The first method was the use of a Sequence of Single-Objective Searches (MOSS) which is based on a reduction in the number of function evaluations reusing the values computed in previous runs. Another method deals with the inclusion of approximate solutions in the NSGA population, that is good genes inoculation into the NSGA population; this genetic modification in the initial population is done using expert knowledge, and using an automatic design alternative like do-nothing, the maximum cost alternatives and full pipe capacity using steady state flow and Manning equation. The third approach for time reduction was to carry out a major modification of the NSGA-II algorithm creating a parallelizing code suitable to be used in a *"Beowulf cluster"* like system; this version is called NSGA-XP. Chapters 4, 5 and 6 describe in detail the optimization and the parallelization process respectively.

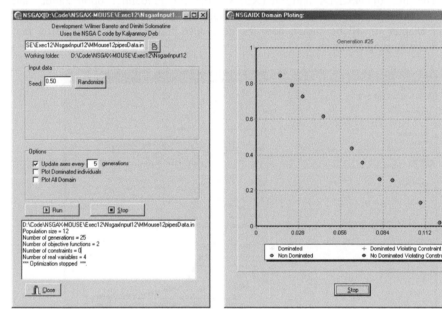

Figure 3.5: NSGAX optimization tool: graphical user interface

3.3.3 Alternatives Reduction

The generated Pareto-optimal solutions are not necessarily feasible or numerically stable, and for this reason a set of rules for post-processing had to be implemented. Another approach is to include these rules in the optimizer of alternatives, penalizing bad solutions; however, it affects the optimization performance due to the extra time needed to check the constraints and also due to the reduction of the feasible domain affecting seriously the optimization performance.

The found Pareto-optimal solutions can be visualized at this stage in OPTRESS; decision makers and stake holders are provided with a suitable graphical iterative tool with the best set of solutions, in which they can negotiate in order to balance the conflicting objectives (Figure 3.6).

3.4 Summary and Conclusions

This chapter showed an approach for urban drainage rehabilitation using a multi-tier framework. The introduced approach is innovative in the introduction of a hydrodynamic modeling inside of a multi-objective optimization process and using parallel computing to for reduction of computational load. It is modular and allows for flexibility depending on the data availability making it suitable for the use in developing countries. It allows for expert knowledge inclusion in different stages of the optimization process. The framework application is simple but without omitting important features for a rehabilitation of drainage systems. The use of NSGAX and OPTRESS tools for modeling, optimization and visualization allows for scalability allowing to expand the tools as much as needed. In the subsequent chapter it will be shown how the developed framework was tested on real problems in developing countries.

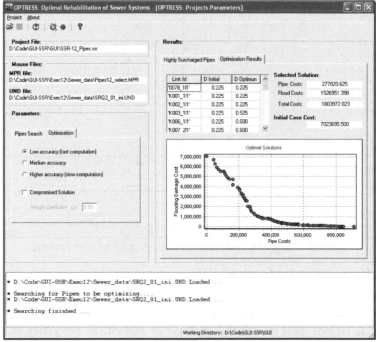

Figure 3.6: OPTRESS tool: viewing the Pareto optimal solutions found by NSGAX tool

Chapter Four

4 Multi-Objective Optimization

First practically usable algorithms to solve optimization problems formally date from 1930s (Leonid Kantorovich) and 1940s (George Dantzig); since then a lot of methodologies and techniques has been developed in the optimization field. A review of optimization techniques to establish the state of the art in the subject is presented in this chapter. Emphasis is made on multi-objective techniques and single-objective global optimization based on population search. An algorithm MOSS specially designed for multi-objective optimization for computationally demanding problems is developed and tested on common benchmarks functions.

4.1 Introduction

For a long-time, engineering processes have been susceptible to optimization. The process to maximize or minimize a procedure is known as operational research. Urban drainage networks can be optimized on cost and performance. In order to understand multi-objective optimization a good background on single optimization is needed first. In this short introduction a brief background on single optimization is given in order to set the scene for multi-objective optimization.

The first step in an optimization process is to define the problem to be solved. There must be a well-defined statement of the problem, including the determination of such factors as objectives, constraints, interrelationships between variables, alternative courses of action etc. This phase is crucial in that the formulation of the problem depends on it.

Secondly, a mathematical model must be formulated. In decision-making, preferences are defined using a payoff function. This function is also called the objective function. The decision-maker's problem can be presented as one of choosing an action among feasible decision variables that maximizes or minimizes the value of this function. The selection of the decision variables can be subject to constraints. A set of models must also be defined representing such constraints.

The decision maker has to transform his problems into a mathematical statement such as:

$$Min_{x} f(x)$$
$$s.t. \tag{40}$$
$$x \in S$$

where *f(x)* is a payoff or objective function, x is a set of decision variables (forming the decision space) that evaluates $f(x)$, x is a vector $(x_1, x_2, x_3, ..., x_n)$ and S is the feasible region in which the selection of solutions is made. This region is defined depending of the constraints of the problem.

After the problem is formulated mathematically, a procedure (method) must be applied to solve the decision-making problem. There are several methods that can be adopted, depending on the nature of the objective function, decision variables and constraints. Usually a good practice is to perform a sensitivity analysis of the variables, to determine critical variables that can affect the solutions.

4.1.1 Traditional Optimization Methods

Several approaches exist to solve optimization problems. They depend on the nature of the objectives, the decision variables and the constraints. Each approach also has a different algorithm (process) to solve it. If the objective function is known analytically, there are several main approaches that are in common use: Linear Programming, Integer Programming, Dynamic Programming, Non Linear Programming and Stochastic Programming.

Linear Programming

Linear programming (LP) is the most widely known tool to solve a certain class of optimization problems. The mathematical equations have the particularity they must be linear in the objective function and in the constraints. A linear programming model is represented as follows:

$$Min \sum_{i=1}^{n} c_i \cdot x_i$$

$$s.t. \tag{41}$$

$$\sum_{i=1}^{n} \sum_{j=1}^{nc} a_{ij} \cdot x_i < b_j$$

Linear Programming is based on four main assumptions (Hillier and Lieberman 2002):

- *Proportionality*: the objective function and constraints equations must be proportional to the decision variables. This means that the exponents in any of the variables must be unity. Also, the use of start-up costs could violate this assumption; Figure 4.1 shows examples of invalid functions.

- *Additivity*: each function in the linear programming model is the sum of the individual contributions of each respective activity. This avoids the cross product of terms.

- *Divisibility*: this applies to decision variables. Since each variable represents the level of some activity, a fraction of it can be used. In other words, these variables are not restricted to be integers.

- *Certainty*: the parameter values: a_{ij}, b_j and c_i, assigned to each function are assumed to be known.

In real problems it is difficult to achieve these assumptions. However, the difficulty can be overcome using a sensitivity analysis. Another option is to change the model. Changes in the model imply changes also in the algorithm used.

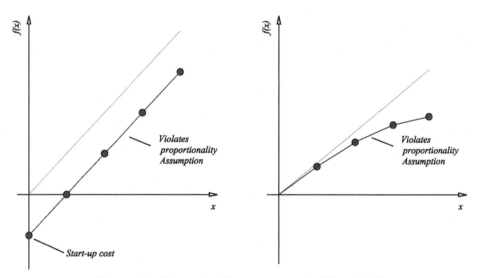

Figure 4.1: Proportionality assumption in LP methods

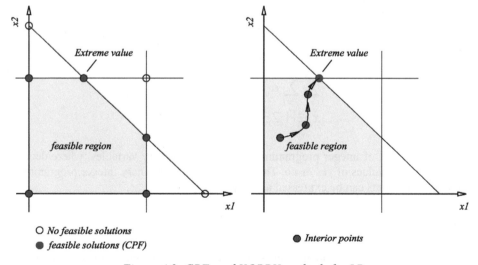

Figure 4.2: CPF and KORBX methods for LP

Linear programming problems are very often solved using the Simplex method suggested by G. Dantzig in 1947. . This is an algebraic procedure with a geometric interpretation. For this reason it is a very powerful method. The method is based on solving the linear equations and finding the corner-point feasible solutions (CPFs). Due to the linearity of the equations the extreme value must be in the CPFs; so their evaluation will give the extreme value as shown in Figure 4.2.

When the number of constraints becomes larger (thousands), the efficiency of simplex methods is reduced. In the 1980s new algorithms have been developed for such cases. One of them is the interior point approach (KORBX). This method starts with a point inside the feasible region, and uses a predictor-corrector method and gradients to compute the next approximation towards the extreme value. Other algorithms such as the dual simplex method, parametric linear programming and upper bound techniques can also be applied.

Integer Programming

There are problems in linear programming for which decision variables cannot be fractions, for example the assignment of people, vehicles or machinery. Also selecting from a list of options is an integer problem. These problems violate the assumption of divisibility. Solving the problem as a linear programming issue and rounding the value to an integer does not necessarily lead to an optimal value.

In this case, the problem must be re-posed including additional constraints on the decision variables. This special kind of problem, in which the decision variables are integers, is called integer programming (IP) and the mathematical formulation can be expressed as follows:

$$Min \sum_{i=1}^{n} c_i \cdot x_i$$

$$s.t.$$

$$\sum_{i=1}^{n} \sum_{j=1}^{nc} a_{ij} \cdot x_i < b_j$$

$$x_i \in \{1,2,3,..., n\}$$

(42)

One particular kind of integer programming is the use of binary variables, where decision variables can take values of yes or no. This approach is called binary integer programming. Additional constraints can be expressed as:

$$x_i = \begin{cases} 0 \\ 1 \end{cases}$$

In most cases fractional and integer values must be used. For instance, to decide whether or not to build a storage facility is a binary decision variable, and the determination of its capacity is a real variable. This kind of model is called Mixed Integer Programming (MIP).

The solution of IP problems involves the simplex method. But the approach is different due to the finite number of solutions that exist. One of the most successful approaches is the Branch and Bounding Technique. This technique is based on dividing the problem into smaller sub problems (branching), bounding the best solution and discarding the sub set if its bound indicates it cannot possibly be a good solution.

There is an alternative to branching and bounding called "*cutting planes*", which can also be used to solve IP. The fundamental idea behind "*cutting planes*" is to add constraints to a LP until the optimal basic feasible solution takes on integer values. Of course, adding constraints has to be done carefully in order to avoid changing the problem. A special type of constraint called a "*cut*" must be added (Figure 4.3). A cut relative to a current fractional solution satisfies the following criteria:

- Every feasible integer solution is feasible for the cut, and

- The current fractional solution is not feasible for the cut.

When it is impractical to compute an optimal solution, one has to settle for a good (but not necessarily optimal) solution. Such solution procedures are called heuristic methods or heuristics. Heuristics often have an intuitive justification, but in many cases they are not guaranteed to produce an optimal solution nor even a good solution.

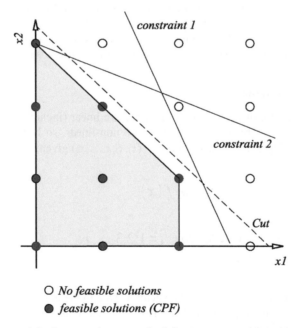

○ *No feasible solutions*
● *feasible solutions (CPF)*

Figure 4.3: Cutting planes method for integer programming

Dynamic Programming

Dynamic programming is a general mathematical technique applied when there is an interrelation between decisions at different stages. There is no standard approach to pose the problem. Therefore, it is not a straightforward task to recognize when to apply DP. However, some guidelines are:

1. The problem can be divided into stages, with a policy decision at each stage.

2. Each stage has a number of states associated with the beginning of each stage.

3. The solution procedure is designed to find an optimal policy for the overall problem.

4. Optimal policies for remaining stages are independent of the policy decisions at previous stages.

5. The procedure starts at the last stage and moves backwards.

6. A recursive relationship that identifies the optimal policy for stage n, given the optimal policy for stage $n+1$, is available.

DP has been used in the solution of graph problems and the allocation of resources. DP can be deterministic or probabilistic. In deterministic DP the state at the next stage is determined by the state and the policy at the current stage. Probabilistic DP includes additional probability terms, which are determined at the current stage.

Non Linear Programming

The main characteristic of LP is that all functions are linear (including the constraints). But in many engineering problems the functions are non-linear, so Non-Linear Programming (NLP) is required. The problem is to find $\mathbf{x}^* = (x_1, x_2, x_3, ..., x_n)$ given:

$$Min_x f(\mathbf{x})$$

$$\begin{aligned} s.t. \\ g_i(\mathbf{x}) \leq b_i \quad for \quad i = 1,2,3,...,m \\ x_i \geq 0 \end{aligned} \tag{43}$$

where $f(x)$ and $g(x)$ are functions of the vector x. Due to the non-linearity of the equations, the feasible region is not a polygon. Also, iso-lines that represent cost function $f(x)$ do not define a plane (see Figure 4.4)

Most NLP algorithms assume the existence of a single extremum and are based on gradient-based descent, however the objective function can have more than one minimum (or maximum). In these cases it is common to use the terms "*local*" and "*global*" optimum. As can be seen in Figure 4.4, a series of local minima can be distinguished along with the global minimum.

Local minima are very important. The methodology used in the optimization process is highly dependent on the existence of local minima in the objective function. Global optimization is the task of finding the global extremum.

An important feature in NLP problems is the kind of "bending". The functions can be concave or convex. Its feature can be determined from calculus using the second derivative:

$$\frac{\partial^2 f}{\partial x^2} \leq 0 \quad for \quad all \quad x \tag{44}$$

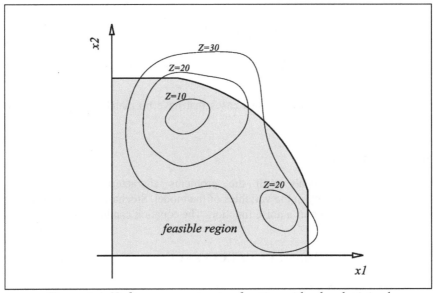

Figure 4.4: No linear programming function with a local minimal

There is no unique method or algorithm to solve a NLP problem. Mathematicians have developed some particular methods, taking advantage of some particular features of the problems. The following methods can be enumerated (Hillier and Lieberman 2002):

- *Unconstrained Optimization*: In this model the function *f(x)* exists, but *g(x)* does not. In such cases if *f(x)* is differentiable *x** can be found using:

$$\frac{\partial f(x)}{\partial x_j} = 0 \quad At \quad x = x^*, \quad for \quad j = 1,2,3,\dots, n \tag{45}$$

- *Linearly Constrained Optimization*: The model is linear in the constraints $g_i(x)$ and its objective function is non-linear. In such cases an extension to the simplex method can be used due to the linearity of the feasible region.

- *Quadratic Programming*: This approach covers a sub-set of linear constrained optimization. Constraints $g_i(x)$ are linear and the objective function $f(x)$ is quadratic. More particular methods have been developed assuming that $f(x)$ is concave.

- *Convex Programming*: This covers special cases where $f(x)$ is concave and $g_i(x)$ is convex. One sub-case is separable programming; in such cases an additional condition must be met: $f(x)$ and $g_i(x)$ must be separable functions. This means each term must involve just one variable x_j.

- *Geometric Programming*: Many engineering problems match this case. The main features are:

$$g(x)=\sum_{i=1}^{n} c_i \cdot P_i(x)$$

$$P_i(x)=X_1^{a^i,1} \cdot X_2^{a^i,2} \cdot X_3^{a^i,3} \cdot, ..., X_n^{a^i,n} \quad for \quad i=1,2,3...n$$

(46)

- *Non-Convex Programming*: All cases that have no match with any of the above methods.

Stochastic optimization

Stochastic optimization deals with the problems characterized by the noise in measurements or uncertainty in the variables of the model. Stochastic optimization is posed as a deterministic function plus a noise function. The equation can be written as:

$$Min_x y(x)= f(x)+\varepsilon(x)$$

(47)

where $y(x)$ is the function to optimize, $f(x)$ is a deterministic term and $\varepsilon(x)$ is the noise term. The presence of noise makes the optimization procedure more complicated, and can lead to undesirable solutions; see Figure 4.5. In such cases more reliable and robust methods are needed.

A special case of stochastic optimization is the use of stochastic programming. This method provides the framework for modeling optimization problems that involve uncertainty (Prékopa 1995). Deterministic optimization problems are formulated with an assumption that all problem parameters are known. Real world problems almost invariably include some unknown or uncertain parameters. When the parameters are known only within certain bounds, the approach is to formulate the problems in terms of probabilities and scenarios. The scenarios have known probabilities and the objective function is minimized for the expected value over all scenarios, and is posed as:

$$Min_x f(x, \varphi)$$

$$s.t. \qquad\qquad\qquad\qquad\qquad (48)$$

$$g(x, \varphi) \leq b_i$$

$$x_j > 0$$

Where φ is a random vector parameter that represents the uncertainties, and is typically associated with a probability.

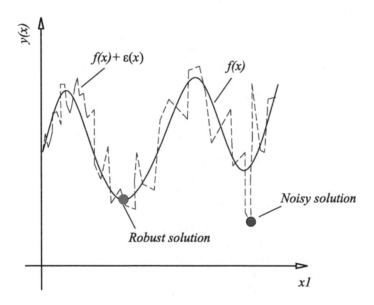

Figure 4.5: Stochastic optimization, robust vs noisy solution

4.1.2 General Optimization Methods

Because most of the optimization problems in engineering and water resources are non-linear, have multiple extreme values (local minima) and most functions may be not be known analytically (so gradients cannot be calculated), other techniques have been developed. Often they are referred to as "global optimization", meaning that the global optimum is sought rather than a local one. The most popular methods are as follow.

Random Search

Random search is one of the oldest methods for global optimization and is a quite straightforward simple algorithm. In it, a number of points are sampled in decision space, the optimized objective function is calculated in each of them, and the sample with the minimum value of the objective function is declared to be the sought solution. It has some advantages such as being easy to code for practitioners. It can also be used for continuous, discrete or mixed functions. However, it is not efficient due to the lack of convergence and its large computational demand due to a large number of function calls. This method is also known as the Monte Carlo technique.

In order to overcome the problems with the number of function calls, and a more efficient covering of the domain space, like Latin Hypercube Sampling can be used. This type of economic sampling distributes a limited number of samples over the decision (search) space (McKay et al. 1979)   but does not ensure the coverage leading to an accurate solution.

Evolutionary computation

In computer science evolutionary computation (EC) is a sub-field of artificial (computational) intelligence involving combinatorial optimization problems. The main characteristics of EC methods are that they are iterative, population based, use guided random search, parallel processing and are biologically inspired. These kinds of methods are also known as "metaheuristics".

The most popular metaheuristic methods are Evolutionary Algorithms (EA), which are inspired by evolution theory, Simulated Annealing and Particle Swarm optimization.

Genetic algorithms (GA) the most popular type of EA, introduced by Holland in 1975 (Holland 1992) and Goldberg (1989). They have been successfully applied in several fields. The method consists in creating a population formed by individuals. Each individual has a chromosome that defines his phenotype given a genotype.

Once a population is created and each individual is coded as a gene; they are ranked using a fitness function. Using this rank and a random procedure (roulette wheel), a crossover is done between the selected parents. The chromosomes are randomly divided and crossed in order to create two new individual and a new population is created. This procedure is repeated until a convergence criterion is met.

Simulated Annealing

Simulated annealing (SA) is a generic probabilistic meta-algorithm for the global optimization problem introduced by S. Kirkpatrick et al (1983), and by Cerny in (1985). The fundamental idea is to allow moves that result in solutions of worse quality than the current solution (uphill moves) in order to escape from the local minima. The probability of doing such a move is decreased during the search.

The name and inspiration come from annealing in metallurgy, a technique involving heating and controlled cooling of a material to increase the size of its crystals and reduce their defects. The heat causes the atoms to become unstuck from their initial positions (a local minimum of the internal energy) and to wander randomly through states of higher energy. The slow cooling gives them more chances of finding configurations with lower internal energy than the initial one.

The algorithm starts by generating an initial solution (either randomly or heuristically constructed) and by initializing the so-called temperature parameter T. Then the following is repeated until the termination condition is satisfied: A solution s' from the neighborhood

$N(s')$ of the solution s is randomly sampled and it is accepted as new current solution depending on $f(s)$, $f(s')$ and T. s' replaces s if $f(s') < f(s)$ or, in case $f(s') > f(s)$, with a probability which is a function of T and the relation $f(s') - f(s)$. The probability is generally computed following the Boltzmann distribution:

$$p(T, f(s'), f(s)) = e^{\frac{(f(s')-f(s))}{T}} \qquad (49)$$

The advantage of this approach is that it is easy to implement, and it always converges to one solution. However, because it is based on "*cooling*" it has a slow convergence rate.

Particle Swarm Optimization

Introduced by Kennedy and Eberhart (1995), Particle Swarm Optimization (PSO) is based on how a bird flock moves towards food. It is one of the heuristic methods based in a population that converge to the optimal in each generation. Particle Swarm Optimization is in a way similar to GA but without genetic operators. It is based on a population that moves in the decision space towards the individual (point) with the best fitness in the current population; this individual or set of coordinates is called "p*best*". The best individual during whole evaluations is also kept in memory, representing the global best fitness so far and is called "g*best*". Each individual moves using a displacement equation, that depends on velocity and acceleration towards these two "*bests*". PSO has been applied in several engineering applications but few in combinatorial problems; see Jarboui et al. (2007), Pongchairerks and Kachitvichyanukul (2009), Izquierdo et al. (2008). The advantage of this method is it uses only a few parameters.

$$v_t + 1 = v_t + c_1 \cdot r_1 \cdot [pbest - x_t] + c_2 \cdot r_2 \cdot [gbest - x_t] \qquad (50)$$

$$x_{t+1} = x_t + v_{t+1} \qquad (51)$$

where:

v	is the velocity vector of the individual
x	is the coordinates vector of the individual
t	is the time step between populations
r_1 and r_2	are random vectors from 0 to 1
c_1 and c_2	are coefficients usually $c_1 = c_2 = 2$

Ant Optimization System

This meta-heuristic method is inspired by how ant colonies behave. They establish communication using a substance called pheromone. From a computational point of view the virtual ants are agents and they interact through pheromone (Dorigo et al. 1996). Initially the method was developed to be applied to graph optimization problems to find the optimal path. Ants travel through different paths using a probabilistic method based in the pheromone associated with that path. If a path has been ranked as good this path has more pheromones and has a better probability of being selected by other ants. This algorithm is suitable to be used on highly constrained complex combinatorial problems (Maniezzo and Roffilli 2008).

Artificial Immune System

Artificial Immune System (AIS) is inspired by how the human body responds to protect itself against bacteria and viruses. The idea is to generate a population of antigen and other antibodies, using conventional GA operators to replicate the better antibodies that match the antigen present (Coello et al 2000). In contrast AIS are able to deal with ever different enemies and so they are flexible and multi-focused. When the organism is exposed to an antigen it produces antibodies to eliminate such an antigen. The production of the antigen is carried out by specialized cells; such cells are also able to keep in memory the antigen in order to produce the specific antibodies when the antigen is present in the organism again. In order to detect a new antigen the cells produce the antibody using random mutations in the composition of the antibodies.

This process can be implemented and used for optimization purposes. In AIS, antigens are the optimal points of an objective function, while antibodies are the test configurations. The optimization search is carried out by modifying antibodies in order to have a better affinity, that is, a greater value of the objective function if a maximum is looked for, to the antigens.

Optimization algorithms based on clustering and multistart

Clustering optimization algorithms are a subset of "multistart" methods. The essence of this approach is to start local search from different initial values. In order to identify the areas where the initial values can be chosen, clustering can be used. A version of multi-start using the downhill simplex descent algorithm by Nelder and Mead (Nelder and Mead 1965) was implemented and tested, for example by Solomatine (1999)

Adaptive cluster covering (ACCO) algorithm was proposed by Solomatine (1995, 1999); in it the local search is replaced by a global adaptive search by random covering associated with the found clusters. It includes the following steps (Figure 4.6):

1. Clustering of "good" solutions: the assumption is that each such cluster will correspond to a region of attraction (where the minimum may be located).

2. Covering: progressively shrinking and moving regions (windows). Random search is used to search for better solutions in each region of attraction, and doing this iteratively, progressively reducing the size of the current region (window) and moving it together with the current best solution (following the direction of the objective function descent).

3. Adaptation: update the algorithm's behavior depending on new information about the objective function obtained after each covering iteration.

4. Periodic randomization: populations can be re-randomized if it is needed to reach the global minimum. This is done due to the probabilistic nature of the random

points generation.

Such a strategy ensures that the general direction of the objective function descent is followed reliably in the search for the global minimum. The number of clusters to be found at Step 1 is a user-defined parameter. ACCO can be complemented with a fifth step: when ACCO has found the global minimum estimate, a local search can be launched (for example, using Powell-Brent or Global Pattern Search algorithms) in order to improve the found solution (this algorithms is referred to as ACCOL). We use ACCO in the MOSS multi-objective optimization algorithm.

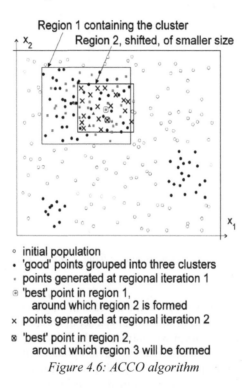

° initial population
• 'good' points grouped into three clusters
· points generated at regional iteration 1
⊚ 'best' point in region 1,
    around which region 2 is formed
× points generated at regional iteration 2

⊠ 'best' point in region 2,
    around which region 3 will be formed

Figure 4.6: ACCO algorithm

4.2 Multi-Objective Optimization

Decision makers have to deal with interests of various parties or stakeholders. Frequently these interests are conflicting and if one objective is met another becomes worse and vice-verse. This kind of problem is approached using multi-objective optimization. Multi-objective optimization consists in optimizing a vector of objective functions, which would give a set of acceptable solutions for decision makers (Osyczka 1984). General Multi-objective Optimization Problems (MOP) can be posed mathematically as: To find the vector $\mathbf{x}^* = [x_1, x_2, x_3, ..., x_n]$ that optimizes the vector function:

$$f(x)=[f_1(x), f_2(x), f_3(x), ..., f_n(x)]$$
$s.t.$
$$g_i(x)>0 \quad for \quad i=1,2,3,...,m$$
$$h_i(x)=0 \quad for \quad i=1,2,3,...,p$$

(52)

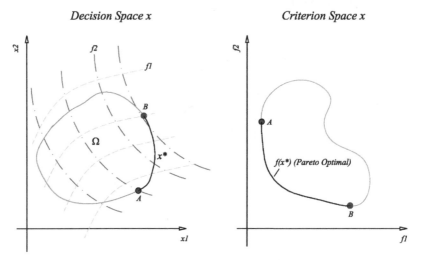

Figure 4.7: Decision and criterion space from a multi-objective perspective

The notion of an optimum in MOP is not like the case of single optimization where just one solution exists. In MOP the concept of optimal is a set of solutions that represents the trade off between objectives. This concept was introduced by Edgeworth in 1881 and generalized by Vilfredo Pareto in 1896. A solution x^* is Pareto optimal if there is no any other x that improves the solution with respect to one objective without making it worse for any other. Figure 4.7 shows a graphical interpretation of the one MOP with two objective functions and two decision variables

4.2.1 Dominance

Dominance is a very important concept in multi-objective optimization. A vector $u=[u_1,u_2,u_3,...,u_k]$ is said to be dominated by the vector $v=[v_1,v_2,v_3,...,v_k]$ if and only if v is less than u in all the components of the vector. It means $v_1<u_1$, $v_2<u_2$, ..., $v_n<u_n$. When at least one component in the vector u is not less than the corresponding in vector v, the vector v does not dominate u. A special dominance concept is weak dominance, and it occurs when the some components in the vectors components are equal.

4.2.2   Traditional methods for MOP solution

Methods to solve MOP are based on the same approach as for single optimization problems. However, the algorithm has to be adapted to find the optimal Pareto solution. In order to solve the Eq. 45 several algorithms have been proposed. Early methods are based on aggregations such as weighted sums, distance to reference objective functions, Keeney-Raiffa method, isovalues curves and the goal attainment method (Collette and Siarry 2003).

4.2.3   Evolutionary Genetic Algorithms

In the last two decades metaheuristic methods have became more prominent. In computer science, evolutionary computation (EC) is a sub-field within the field of artificial intelligence involving combinatorial optimization problems. The main characteristics of EC methods are that they are iterative, population based, use guided random search, parallel processing and are biologically inspired as stated previously. These kinds of methods are also known as "metaheuristics" and by different authors are typically referred to as Evolutionary Genetic Algorithms (EGA), Evolutionary Algorithms (EA) or Genetic Algorithms (GA). There are several multi-objective GAs and the most used ones are: Non-Dominated Sorting Genetic Algorithm (NSGA-II) and Epsilon Multi Evolutionary Algorithm (ε-MOEA).

NSGA-II Agorithm

NSGA-II algorithm was developed by Deb et al. (2002) as improvement of its antecessor NSGA also developed by the same author. In NGSA-II the offspring population Qt is first created by using the parent population Pt, based on mating and mutation processes, these two populations are combined forming a new population of $2N$ dimension (Rt). Then a non-dominated sorting is used to classify the entire population $Rt;$ this classification is made using a ranking (fitness) function to identify several fronts ($F1,F2..Fn$).

Due to the new population being greater than N, just the best rank individuals remain; in this way the elitism is kept. Usually more than N individuals have the same ranking (belong to the same front), individuals are not deleted arbitrarily, they are selected in such way that they reside in the least crowded region using the concept of crowding distance; this ensures diversity. Figure 4.8 summarizes the selection process.

Epsilon-MOEA

Epsilon-MOEA, introduced by Laumanns (2002) and Deb (2003), is based in the concept of ε-dominance, which does not allow for two solutions with a difference less than ε in each objective function to dominate one another. The search space is divided into cells and the diversity is maintained by ensuring only one solution per cell. The convergence (elitism) is preserved using an archive that contains the all ε-non-dominated solutions. A normal individual from the population is selected by a tournament (based on the dominance concept) and is combined with a randomly selected parent from the archive; the child is compared with each individual from the archive and if it is ε-non-dominated by them it becomes part of the archive. Figure 4.9 shows the schematic diagram of Deb Epsilon-MOEA.

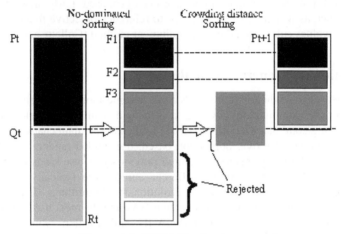

Figure 4.8: NSGA-II selection procedure

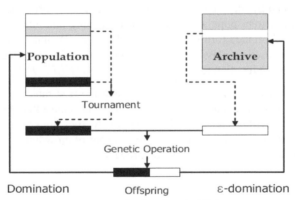

Figure 4.9: Epsilon-MOEA

4.2.4 Performance and Pareto Comparison

One of the main problems in multi-objective optimization problems is establishing the stopping condition. Also, the comparison between Pareto sets is an issue. In order to establish a suitable stopping condition and allow comparison between Pareto sets a suitable metric indicator has to be selected.

There are several possible metric indicators in order to compare Pareto sets. However as stated by Zitzler et al (2002) that there is no unique quantitative indicator that can tell us which one of the Pareto sets is better, and it is not always possible to decide about the advantages or disadvantages of such indicators. Five traditionally used indicators have been selected for this study: cardinality, number of function calls, computational time, hyper-volume and ε -indicator.

Cardinality is the number of solution vectors in the Pareto set, which is an indirect measurement of the diversity. The number of function calls and the corresponding computational time is of extreme importance in urban drainage due to the long running time of the underlying models. Hyper-volume is a measure of the hyper-space dominated by the Pareto front; it is computed by slicing one objective at time and computing the hyper-space (Figure 4.10). The ε–indicator ($I\varepsilon$) is a scale factor that indicates how much a Pareto set "A" must be moved or scaled in order to dominate a Pareto set "B" (also called Pareto set of reference); see Eq.44.

$$I_{\varepsilon}(A, B) = \underset{z^2 \in B}{Max} \quad \underset{z^1 \in A}{Min} \quad \underset{1 \leq i \leq n}{Max} \frac{Z_i^1}{Z_i^2} \qquad (53)$$

where: $I\varepsilon$ is the epsilon indicator. A and B are the Pareto sets to compare, z^1 and z^2 are the objective function vectors of A and B respectively and n is the number of objective functions.

4.3 Multi-objective Optimization for Computationally Demanding Problems

Multi-Objective optimization has become an important tool for engineers and decision makers. Being known for a longer time, it has not been widely used in engineering "*real*" problems due to the lack of suitable methods to deal with complex problems. Usually, problems in engineering are solved using numerical methods or stochastic analysis which are time and computationally demanding, taking minutes to hours for a single run. Such problems can be also referred as "*expensive*" problems. In this case reducing the number of function evaluations and finding "at least some good solutions" becomes the main problem, and it can outweigh the traditionally used criteria to evaluate the quality of multi-objective optimization algorithms (diversity, cardinality, etc.).

The development of global optimization approaches combined with heuristic search, like Evolutionary Algorithms (EA), has facilitated the evaluation of complex function. No derivatives or additional knowledge of the functions is required. However, such algorithms often require many function evaluations, and this may make some of them prohibitively time-consuming.

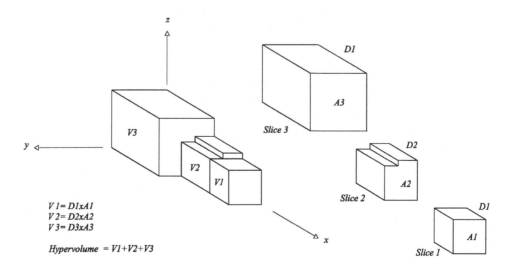

Figure 4.10: Hyper-volume computation using slicing method

This section addresses such computationally demanding problems. Water distribution and drainage networks are good examples of these (one model run may require several minutes or even hours). Successful application of multi-objective optimization has been reported for such kind of problems; however, it just has been applied to low-dimensional problems. Relatively little attention has been paid to multi-objective optimization of problems in which the number of function evaluations is critical; see Knowles (2006), Vamvakeridou-Lyroudia et al. (2005), Barreto et al. (2006). Most of the work is oriented to the use of the approximating (surrogate) meta-models in order to reduce function evaluations in the optimization. The use of meta-models based on artificial neural networks (ANN), response surface approximation and use of kriging based models in Gaussian processes is reported by Broad et al. (2005), Goel et al. (2007) and Knowles and Hughes (2005).

Each of these methods has advantages and disadvantages. In the case of modeling using ANN, the gain could be marginal if the training and validation sets (function calls), are taken into account in order to build the surrogate model. The application of Gaussian models reported by Knowles using ParEGO looks very promising but has been tested only for low-dimensional problems.

There is however an alternative way to reduce the number of function calls – to use the scalar functions. Scalar functions to build the Pareto front were already used in muticriterial optimization research years ago. Aggregations like weighted sums, distance to reference objective function, Keeney-Raiffa method, isovalues curves and goal attainment method can be mentioned; see (Collette and Siarry 2003) for further explanation of each method. Aggregation approaches are easy to implement and computationally efficient but they are not able to find more than one point in the Pareto set. In order to find a diverse

Pareto set, the scalar problem must be solved iteratively for each preset weights values.

The above approach makes the procedure more complex than to solve the problem using a multi-objective evolutionary algorithm MOEA. Hajela and Lin (1992) encode the weights in the chromosome of a genetic algorithm (GA) to deal with the diversity problem in the Pareto set. (Jin et al. 2001) presented a GA approach applying weighted sum, changing the weights of each objective in a fixed number of iterations. Both approaches solved the diversity problem but still require a high number of evaluations.

4.3.1   Multi-Objective Optimization by a Sequence of Single-objective Searches (MOSS algorithm)

This work introduces an approach to multi-objective optimization by a sequence of single-objective searches (MOSS). It makes it possible to deal with expensive highly dimensional problems, which are time consuming, and their high number of variables makes the convergence to an extreme value problematic. It is based on the idea that in most engineering problems the whole Pareto set is not needed, and practitioners are typically satisfied with a limited number of solutions belonging to this set. The computation of optimal solutions, over the Pareto set, can be restricted to a decision maker preference zone, reducing in this way the feasible region. In such cases, the use of a few (3 or 5) scalar weighted functions are suitable to represent the interest area and thus reduce the number of functions evaluations for expensive problems. Note that in such a setting the criteria of diversity and cardinality become much less important for a decision maker whose desire is to obtain only a small number of reasonably good Pareto solutions.

In the MOSS framework, for optimization of the aggregated function any randomized search algorithm can be used, for example a GA. In the past, we had a positive experience with the Adaptive Clustering Algorithm (ACCO) (Solomatine 1995, 1999, 2005). This algorithm often outperforms several multi-start local search algorithms and canonical GA in real cases problems, as is shown by Abebe and D.P. Solomatine (1998). In order to evaluate the performance of the method, four benchmark functions have been evaluated. For the purpose of the qualitative comparison with MOSS, the well-known NSGA-II (Deb et al. 2002) has been applied to the same set of problems as well.

Distance to a Reference Objective

Weighted aggregation methods for multi-criteria optimization are well documented. The most common approaches are: weighted sum (Eq. 40) and distance to a reference objective vector (Eq.43).

$$Min \quad f_{eq}(\boldsymbol{x}) = \sum_{i=1}^{k} w_i \cdot f_i(\boldsymbol{x}) \tag{54}$$

$$Min \ L[f(\boldsymbol{x})] = \left[\sum_{i=1}^{k} |w_i \cdot (\boldsymbol{F}_i - f_i(\boldsymbol{x}))|^r \right]^{\frac{1}{r}} \tag{55}$$

where:

k	is the number of objective functions
$\boldsymbol{x}$	is variable vector
f_i	is the ith objective
$\boldsymbol{F}_i$	is the ith reference objective
w_i	is the applied weight to an objective i^{th}
r	is a value bounded on $1 \leq r \leq \infty$

These equations correspond to isovalue curves. In the case of a weighted sum, this isovalue curve corresponds to a straight line (Figure 4.11a). The weighted distance to a reference objective produces a conic isovalue curve, circular or elliptical, depending on the weight values (Figure 4.11b).

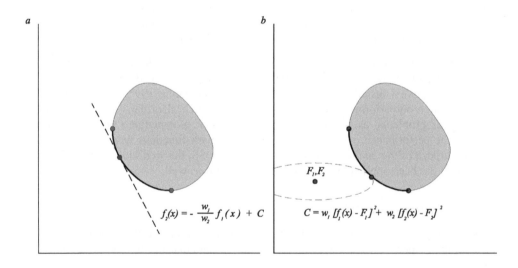

Figure 4.11: Isovalue curves, 2(a) Weighted sum and 2(b)Weighted distance to a given reference objective set

One of the main limitation of these methods is that typically they are not able to deal with non-convex regions, despite the fact that some authors have shown they could work for convex and concave Pareto sets (Jin 2001). In order to improve the ability of the non-convex search, the distance function in a reference objective method was modified. The weighted distance to a reference objective method makes an anisotropy search, giving more importance to one axis direction. With the inclusion of an extra degree of freedom in the equation, namely rotation, it is possible to search with a preferred direction different from the main axes (Figure 4.12). In the case of a bi-objective problem and using Euclidean distance ($r=2$) the canonical equation without rotation looks like Eq.42. Using rotation a new term appears which represents the rotated isovalue curve (Eq. 41).

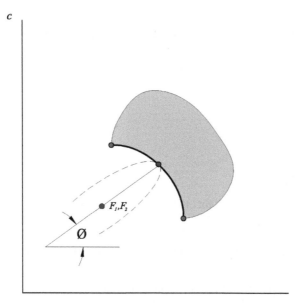

Figure 4.12: Isovalue curve rotation using weighted euclidean distance

$$L[f(x)]=\left[A\cdot(f_1(x))^2+B\cdot(f_2(x))^2\right]^{\frac{1}{2}} \tag{56}$$

$$L[f(x)]=\left[A\cdot(f_1(x))^2+B\cdot(f_2(x))^2+C\cdot f_1(x)\cdot f_2(x)\right]^{\frac{1}{2}} \tag{57}$$

where:

A,B and C	depends of **F** and the rotation angle φ
f_1 and f_2	are the objective functions
x	is variable vector
L	is the euclidean distance

All the described methods were implemented, but in the presented case studies only the distance method was used.

Single-objective optimizer: Adaptive Cluster Covering Algorithm (ACCO)

MOOS can use any single-objective optimizer, but we used Adaptive Clustering Covering (ACCO) as introduced by Solomatine (1995, 1999) and described above. It was chosen to the attractive feature: quite a low number of function calls needed to arrive to a good estimate of the minimum, and the possibility to save the user-defined number of potentially good points in a flexible way. For the use in MOSS, ACCO was modified as described below.

MO Optimization Using a Sequence of Single-Objective Searches (MOSS)

Consider a problem where a decision maker desires to find M solutions in the Pareto set. The following algorithmic scheme is proposed:

1. Archive $S := empty$

2. **for** $m := 1$ to M **do**

   begin

3. Generate set of weights $wm = [wm_1 .. wm_k]$

4. Construct function f_{eq} (Eqs 42 or 41) using wm and f_i

5. Run single-objective minimization algorithm for f_{eq} and while doing so:

   a. Use points and the corresponding calculated f_i from archive S whenever possible

   b. While performing minimization, add all generated points in decision space together with the calculated values f_i to archive S

   end;

In Step 5 the modified ACCO (or ACCOL) algorithm is used. The modification concerns the following: the generated points together with the corresponding "precious" calculated function values f_i, are saved to an archive in order to be used if possible, during the subsequent minimization runs for different sets of weights wm. The motivation of such a modification is to try to reduce the number of the needed function evaluations. ACCO was chosen for this role because its software implementation allows for the subsequent use of any (user defined) number of solutions generated by its previous runs. The tests have shown that this approach works well. Figure 4.16 illustrates this approach. This algorithmic scheme will be referred to as MOSS/ACCO or MOSS/ACCOL depending on the algorithm used for single-objective optimization.

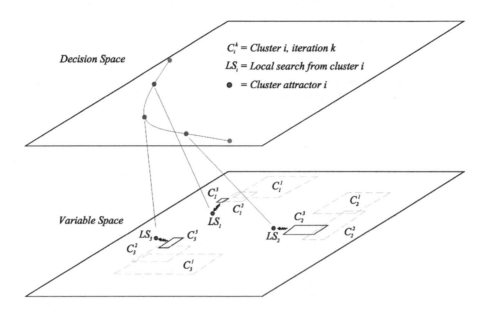

Figure 4.13: Use of ACCO algorithm in multi-objective optimization

Comparing Algorithms Criterion

For computationally expensive optimization problems, the criteria traditionally used to compare multi-objective optimization algorithms (diversity, cardinality, proximity to a reference Pareto set, etc.) should be complemented by yet another criterion (and often a more important one) the number of function evaluations. A decision maker often has to be satisfied with only a small number of Pareto solutions since finding more is prohibitively expensive, so the criteria of diversity and cardinality become less important. A possibility is to move some evaluation criteria to constraints and to compare the algorithms based only on one criterion, for example on a minimum number of function evaluations given a certain cardinality, diversity, or on maximum diversity given a limited number of function evaluations and cardinality.

The comparison of different algorithms will be based on how fast (measured by the number of functions evaluations) an algorithm identifies Pareto solutions (with some constraints on the cardinality or diversity) that are close to a reference Pareto set.

We understand the limitations of such a comparison, and clearly see the necessity of developing a proper methodology for comparing different algorithms for expensive problems. The complexity here is that various algorithms were developed for different purposes, often trying to satisfy some of the mentioned traditional criteria. For example, most existing MOEAs were optimized with respect to diversity, cardinality and proximity to a reference set, and not for the number of function evaluations, so comparison of our approach to them could be only qualitative, if justified at all. In this thesis, along with the results achieved by our algorithm, the results obtained with NSGA-II are presented as well, but it would not be right to perform a quantitative comparison due to differences in the principles laid out in the design of both algorithms.

Test Problems

Four well-known multi-objective benchmark functions (Deb et al. 2002) where selected to test the approach (Table 9). Both convex and non-convex functions of low and high dimensions were selected.

Table 9: Benchmark function and weights

Function	Variables	Bounds	Weight Vector	Clusters	Comments
SCH	1	[-1000,1000]	[0.7 0.3; 0.5 0.5; 0.3 0.7]	1	Convex
ZDT1	30	[0,1]	[0.7 0.3; 0.5 0.5; 0.3 0.7]	3	Convex
ZDT2	30	[0,1]	[0.7 0.3; 0.5 0.5; 0.3 0.7]	3	Non Convex
ZDT4	10	X_1 [0,1] X_i [-5,5]	[0.7 0.3; 0.5 0.5; 0.3 0.7]	4	Non Convex

In the presented algorithmic approach, three Pareto solutions (corresponding to the three different weight vectors used in Eq. 40) were sought for each function. The number of ACCO clusters per solution ranged from 1 to 4, depending of the number of variables. Table 9 shows the weight vector used for each function and the cluster number that was used.

For comparison purposes, the Pareto front was also found using the NSGA-II algorithm. Each function was run 25000 times, and each generation was saved in order to compare with MOSS (using ACCOL as the single-objective optimizer) for various numbers of function evaluations.

For the SCH and ZDT1 functions, MOSS performed much better than NSGA-II. It was able to achieve the reference Pareto front in 240-function calls, while NSGA reached the reference Pareto after 400 functions calls for SCH. For ZDT1, MOSS needed 2800 and NSGA-II around 5000 function calls. In the case of ZDT2 function, MOSS performed marginally better than NGSA-II with 3800 function calls. For ZDT4 function, MOSS was not able to find a solution close to the solutions in the reference Pareto set. Figure 4.14 to 4.15 shows MOSS vs. NSGA-II comparison for 3 benchmark functions. These preliminary results make it possible to conclude that the approach is promising. As described above, an accurate comparison of its performance to that of other algorithms is hardly possible due to the different principles laid out in the design.

Except for the function ZDT4, in which NSGA-II outperforms MOSS, it was performing better in terms of the number of function evaluations than NSGA-II to converge to a solution over the Pareto, saving around 40% of the function evaluations for 2 different functions (SCH and ZDT1), and 16% for ZDT2. Also, it was possible to identify other non-dominated points in the reference Pareto set, giving a better diversity in the solutions. However, the degree of confidence in such "extra" values has to be explored.

4.4 Summary and Conclusions

A brief review of the state of the art for optimization in engineering problems was done. It showed that there is no unique method for optimizing all kind of problems, but it is necessary to choose a method fitting the problem at hand. In the area of urban drainage very little examples of large scale optimization could be found, and practically no applications to real cases using hydrodynamic models inside of the optimization algorithm.

A special type of problems were identified as "computationally demanding" (or "expensive") problems. Multi-criteria urban drainage rehabilitation falls on the ambit of such kind of problems, where high computer power is required. A Multi-objective Optimization algorithm using a Sequence of Single-Objective Searches (MOSS) to face these is developed and tested using four benchmark functions for multi-objective algorithms. The method is based on a pragmatic idea that practitioners typically need only a few solutions close to Pareto optimal set rather than a large set of them most of which are similar. The new approach outperforms NSGA-II on three of the benchmarks, while NSGA-II was slightly better for the fourth benchmark functions .

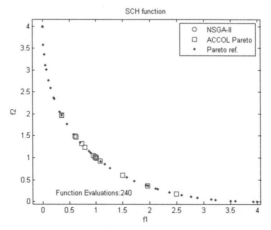

Figure 4.14: SCH function. After 240 iterations MOSS identified several solutions very close to the reference Pareto set. NSGA-II solutions are still too far to appear in the plot range

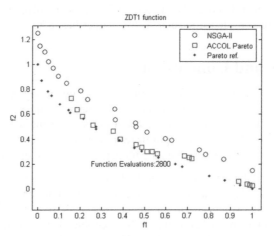

*Figure 4.15: : ZDT1 function. MOSS and NSGA-II
results after 2800 iterations*

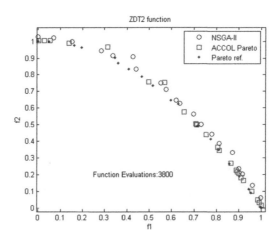

*Figure 4.16: ZDT2 function. MOSS and NSGA-II
results after 3800 iterations*

Chapter Five

5 Multi-Objective Optimization in UDS Rehabilitation

This Chapter shows the implementation of the multi-tier framework described on Chapter 3 in a prototype. It can be seen as a proof of concept on the applicability of the framework. Urban drainage rehabilitation investment costs measures are reviewed and implemented, pipe renewal, storage tanks and ponds and diversions are implemented and tested in a small network in Denmark. Damages cost are also implemented. Storage ponds and tanks are evaluated in a case study. Two multi-objective genetic algorithms are tested and checked using four performance indicators.

5.1 Introduction

In order to optimize the budget and other objectives in an urban drainage rehabilitation plan it is necessary to build appropriated objective functions; they can be expressed as monetary or non-monetary cost functions. One of the most important objective function to decision-makers is the construction costs; it is associated with many variables that affect mainly structural quantities and consist in the replacement or extension of parts of the drainage network. Construction costs are manly related to renewal or repair of pipes and channels, manholes, additional storage, land costs, control structures, pump systems, infrastructure, among others. In order to build the cost functions a cost related to maintenance has to be also included in such a cost function. In this chapter two main rehabilitation methods are evaluated and implemented within the framework of the present research: pipe renewal and storage. These have to take into account costs of pipes, excavation works, placement of pipes and reinstatement of the ground surface; this has to be done through unitary cost analysis for the pipe and storage.

In addition to construction costs, the damage caused by flooding has to be taken into account. This is treated as another objective in order to contrast inversion costs and flood damages in the formed Pareto. Damage costs can be evaluated in several ways, the most common approach is to take into account the land use and damages depth curves in the inundated area. Also, water velocity can be used for the flood damage estimation; water depth and velocities are considered using a hydrodynamic model and modeling the above- and below-ground drainage structures. Other objective functions, as indirect damage costs or intangible costs are also implemented and tested in this chapter.

5.2 Pipe renewal in UDS Rehabilitation

Replacement of pipes may be required for several reasons. The main reason is that a pipe has reached its service life and the maintenance cost becomes higher than the replacement cost. Also, when a pipe is not providing the expected service it may be required to be changed. When a pipe is broken or collapsed decision-makers have to decide whether or not the pipe must be replaced. This is typically done through cost benefit analysis.

There is no a universally acceptable comprehensive method which indicates when to change a pipe. Most of the methods that are used for this purpose can be classified into the following four categories (Loganathan et al. 2002).

- Deterioration point assignment (DPA) schemes: these start with the identification of the intervening variables which affect pipe failures; for instance, the diameter, length, aging, and pipe material, type of soil, and historical breaks of pipes, among others. These variables are numerically scored in rank, weighted and added for grand total. If the computed score exceeds a fixed threshold the pipe is a candidate to be replaced or repaired.

- Breakeven analysis: these methods are based on a cost approach; an amount of money has to be deposited at a fixed interest rate, and this amount is compared with pipe costs of replacement/repair to be incurred in the future. These methods need a pipe failure forecast analysis.

- Regression and failure probability methods: these are methods that take into account the same variables and look much like the DPA methods but they have predictive capabilities by assessing the probability of survival. Regression analysis establishes a relationship between the rate of breaks in a pipe and the time; this regression is used to find the time which minimizes the replacement/repair costs (Shamir and Howard 1979). This minimization can also be related to the frequency of breaks in the pipe, in other words, by finding a threshold based on the number of breaks per year (Walski and Pelliccia 1982).

- Mechanistic strength assessment methods: These are based on regression or physically based models to simulate the failure mechanisms of a pipe. They are able to introduce other variables like hydraulic loads, corrosion, temperature, stress, traffic loads, among others. Also, they use probabilistic models due to the complexities and randomness of the variables. Until now, this kind of model is not very well developed and imply much effort in its application.

In developing countries, where existing data is scarce and difficult or costly to acquire, the methods to be used must be simple, in order to take the decision whether or not to change a pipe. This does not imply that the same approach must be taken in the future; in the meanwhile, data has to be improved and collected for a more refined approach. Based on this requirement the categories which appear most suitable are DPA schemes or Breakeven analysis.

5.2.1 Pipe renewal selection

Before starting the optimization process it is necessary to select the pipes which are candidates to be renewed. Two criteria are used in pipe selection: one criterion is related to surcharge in pipes which is a hydraulic criterion, and the other can be related to the present worth of the pipe. The hydraulic criteria is based, as explained in Chapter 2, on the relation between the real discharge and full pipe capacity discharge; in other words the Q/Qf ratio.

The use of the total present worth can help the selection of whether or not a pipe is suitable to change. It is done through the minimization of the present worth. In order to compute the present worth it is necessary to estimate the maintenance and replacement costs of the pipe; they are related to several factors, the most relevant being the number of breaks and the age of the pipe. If we assume that a given pipe will be changed at a time n years of its expected life, its total present worth can be expressed as:

$$T_n = \sum_{i=1}^{n} \frac{C_i}{(1+r)^i} + \frac{F_n}{(1+r)^n} \tag{58}$$

where:
$\quad T_n \quad$ is the total present worth

$\quad n \quad$ is the year in which is expected to change the pipe

$\quad i \quad$ is the year i^{th}

$\quad C_i \quad$ is the pipe repair cost at year i^{th}

$\quad F_n \quad$ is the pipe replacement cost at year n^{th}

$\quad r \quad$ is the discount rate

When the pipe is new, few breaks are expected giving a relatively low component due to repairs; but if the pipe is changed at an early stage, the replacement cost is high. In contrast, changing the pipe at a later stage will reduce the replacement cost. However, the number of breaks in older pipes are higher than in new ones, thus increasing the repair cost of the pipe. As illustrated in Table 10 column 2 shows the history of breaks for a pipe. If the repair cost per break is € 1,000 and the replacement cost is € 150,000, the total present worth (TPW) of the pipe can be calculated as shown in column 6 using a discount rate of 7.5%. As can be seen, the TPW starts at a high value and then decreases, until it reaches a minimum value of € 128,340 at year 5; then it starts to grow again due to the number of breaks until it reaches a critical value; see Figure 5.1. A minimum value of TPW of € 128,340 can be identified at year 5; beyond this year the pipe can be considered as a candidate for change.

Table 10: TPW and EUAC for a given pipe

		Costs in €				
year	Breaks	$C_i/(1+r)^i$	Repair	Replace	TPW	EUAC (€/yr)
1	2	1,860	1,860	139,535	141,395	152,000
2	4	3,461	5,322	129,800	135,122	75,253
3	6	4,830	10,152	120,744	130,896	50,334
4	9	6,739	16,891	112,320	129,211	38,578
5	10	6,966	23,856	104,484	128,340	31,721
6	13	8,423	32,280	97,194	129,474	27,584
7	14	8,439	40,718	90,413	131,132	24,758
8	16	8,971	49,690	84,105	133,795	22,842
9	19	9,910	59,600	78,238	137,837	21,608
10	21	10,189	69,789	72,779	142,568	20,770
11	25	11,284	81,072	67,701	148,774	20,337
12	29	12,176	93,248	62,978	156,226	20,197
13	33	12,889	106,137	58,584	164,721	20,271
14	36	13,079	119,216	54,497	173,713	20,463
15	40	13,519	132,735	50,695	183,430	20,780
16	44	13,833	146,568	47,158	193,726	21,192
17	50	14,623	161,190	43,868	205,058	21,736
18	50	13,602	174,793	40,807	215,600	22,213
19	59	14,931	189,724	37,960	227,684	22,862
20	63	14,831	204,555	35,312	239,867	23529

Total present worth has the tendency to find the optimal changing value in the first quarter of the expected pipe life; even if intangible or social costs are included the time in which a pipe has to be changed is less (Deb et al. 2009). Another methodology that can be used is the equivalent uniform annual cost (EUAC). EUAC is preferred when the expected lives of each asset in the system are different, while the TPW is preferred for the same life time of the network assets (Deb 2002). For the estimation of EUAC it is necessary compute the TPW first, and use it to find a uniform distribution of cash flow; it can be compared with the payments for a loan at a given discount rate:

$$EUAC_n = TPW_n \frac{r(1+r)^n}{((1+r)^n - 1)} \tag{59}$$

where: TPW_n is the total present worth at year n^{th}

                 n is the year in which is expected to change the pipe

                 $EUAC_n$ is the equivalent uniform annual cost at year n^{th}

                 r is the discount rate

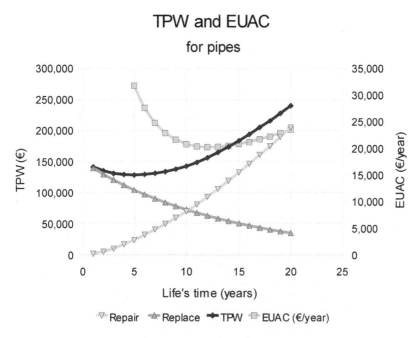

Figure 5.1: Total Present Worth and EUAC minimum values

As can be seen from the Table 14 and Figure 5.1, EUAC starts with a high value and decreases until it reaches the minimum value at year 12; after that it starts to grow slowly. It can be inferred that a pipe after 12 years is a candidate to be changed. As can be seen TPW is more conservative than EUAC at the moment of taking the decision whether or not a pipe is suitable for change. EUAC appears to be more realistic than TPW.

The selection of the pipe is done according to two rules or thresholds; the first one is the hydraulic condition which is evaluated using the ratio between the discharge in the pipe and the pipe capacity, and the other is the age of the pipe taking into account the PTW or EUAC. This can be summarized by:

$$Select\ pipe\ if \begin{cases} \dfrac{Q}{Qf} > Q_{threshold} \\ \qquad or \\ \dfrac{Age}{Age_{opt}} > Age_{threshold} \end{cases} \tag{60}$$

where:

	Q	is the total pipe discharge
	Qf	is the full capacity pipe discharge
	$Q_{threshold}$	is the selected thresholds of pipe rate discharge
	Age	is the current pipe age
	Age_{opt}	is the optimum time for pipe change
	$Age_{threshold}$	is the selected thresholds of pipe rate age

5.3 Pipe renewal Cost

In order to carry out a pipe optimization it is necessary to define the cost function to evaluate. The first step is to define the variables that affect the total cost for a pipe installation; variables such as diameter, length and slope of the pipe are important. Also, the flow which will be carried, friction factor, pump systems efficiency and power have influence on the pipe cost. Some of these variables are known beforehand when we talk about rehabilitating pipes. For example, length, slope, materials or friction factors are known. In this application it is assumed that there is no pump system; all this leads us to the conclusion that only diameter and length of the pipes are influencing costs.

Two costs are the most important in pipe renewal; one is the cost of the pipe itself, including the installation (or placement) costs, and the other is the cost of the trench excavation. The second item is variable depending on factors such as machinery to be used, type of soil, land use , etc. For the trench excavation it can be assumed that on average they are buried at 2 m with a width of the trench of 60 cm plus the pipe diameter. The fixed cost, and special cost can be estimated as 10% to 15% of the pipeline costs (Swamee and Sharma 2008). Using this simplified analysis or a more complex one, which includes the costs of materials, equipment and labor, plus administration cost and tax for each diameter leads to the development a catalogue of pipe costs. This catalogue can be formulated as:

$$C = k \cdot D^m \tag{61}$$

where:

	C	is the pipe cost per unit length
	k	is the a coefficient

$$m \qquad \text{is the an exponent}$$

$$D \qquad \text{is the pipe diameter}$$

Table 11 shows the catalog used in this research. It provides the basic input for the cost function for a pipe renewal project. In Figure 5.21 the parameters of the Eq 61 have been selected; it can be seen that k is 0.56 and m is 1.17. The developed software has been set to use a tabular catalogue instead of Eq 61; this is because it is not possible all the time to obtain only one slope when values are plotted on log-log axes; this leads to the use of more than one equation.

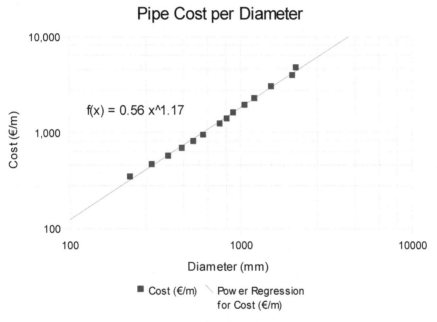

Figure 5.2: Pipe cost fitting to Eq 61

Inflation effects

Developing countries are used to having high values of inflation. The inflation is a rise in the price of services or goods over a period of time. In general terms, if the inflation is taken into account in a project, the present worth will be grater than the actual investment for a particular good, or in our case the pipe network. This is unacceptable from an engineering point of view. However, if the idea is to compare projects or alternatives over the same period of time, inflation has not much influence in the selection, because all the alternatives are affected by the same value. In this thesis the effect of inflation is omitted for the reasons explained above.

Table 11: Catalog of pipe costs per unit length

#	D(mm)	Cost (€/m)
1	225	350
2	300	470
3	375	575
4	450	692
5	525	818
6	600	954
7	750	1,252
8	825	1,414
9	900	1,632
10	1050	1,955
11	1200	2,301
12	1500	3,058
13	2000	4,000
14	2100	4,807

5.3.1 Objective function for pipe renewal

Once the catalogue is built the objective function for pipe renewal can be written as a function of the length of the pipe:

$$C_p = \sum_{i=1}^{n} \left[C(D_i) \cdot L_i + AD_i + P_i \right] \tag{62}$$

where:

	C_p	is the total pipe cost
	i	is the index for pipe i^{th}
	n	is the number of pipe to be changed
	D_i	is the pipe diameter i^{th}
	$C(D_i)$	is the cost for the pipe i^{th}
	AD_i	is the residual value of the pipe i^{th}
	P_i	is a penalty function

In Eq 62 there are two additional factors that are included in the cost function. One is the AD_i term; this term represents the residual value of the pipe. If a recently installed pipe, with an age less than the optimal age to be changed, has been selected as a candidate for change, its total cost will be greater than one which has reached its life cycle. The method assumes that the residual cost is zero in the year in which the pipe has to be changed; it can be estimated using TPW or EUAC methods. This *AD* value is a fixed value, it means that during the optimization process it will not change at all; however, it has an influence on the optimization regarding the selection of one pipe or another.

The second term that has been included is the P_i term; this is a penalty function. There are some diameter values that may incur a violation of some rules. For instance, in a drainage network it is expected that diameters increase down-stream. However there may exist a solution which can reduce the cost using a smaller diameter down-stream; in such a case the penalty function can be activated to avoid this situation. This feature has to be used carefully, because it could reduce the decision domain.

5.4 Storage in UDS Rehabilitation

Another option that can be included in a rehabilitation project is storage. There are two main types of storage that are typically considered: dry and wet storages. Storage tanks or ponds are aimed at reducing the peak runoff during an event. They retain the water during the event and release it at slower rate. Also, they can be used as treatment plants to eliminate some contaminants from the water. The storage can be classified in tanks and ponds. Storage tanks in cities with a high density of inhabitants are usually located underground and made of reinforced concrete, while ponds are built on the surface. Ponds can be classified as dry and wet detention ponds.

5.4.1 Dry and wet detention ponds design

Dry ponds can be used almost under any conditions to reduce peak flows; they are aimed at reducing peaks but they do not perform well when pollutant controls become important. If they are well designed with a retention time greater than 24 hours, they are able to remove part of the pollutants through the settling of particles, but they will not remove the dissolved pollutants (US-EPA 2011), (Scholes et al. 2008) . Dry ponds reduce peaks but do not reduce the total volume of runoff, and this causes problems in downstream areas.

Wet ponds are similar to dry ponds, but they retain a volume of water during the whole year. They require more surface area than dry ponds because of the additional volume. They have better capabilities regarding pollutant removal. In arid zones they are not recommended because of the need to add water to maintain a permanent water volume, which raises the maintenance costs. Also, the geology is important to avoid high losses due to infiltration.

In order to establish costs it is necessary to know how the design of the ponds has to be done. In general terms, the pond has to be designed to give protection for a peak discharge of 2, 10, and 25 years of return period; also, it also has to be able provide attenuation for 100 year event. A detention pond has to consist of the following elements:

- An outlet structure

- An spillway structure for emergency

- •   Maintenance Access and,

- •   A good landscaping design

Figures 5.3 and 5.4 show the typical elements for dry and wet ponds.

Another important issue related to the costs of dry or wet ponds is their maintenance. A comprehensive review about maintenance can be found at WMI (1997). Table 12 shows the activity maintenance and time interval of such activity.

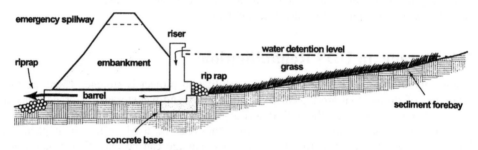

Figure 5.3: Dry pond design (USEPA)

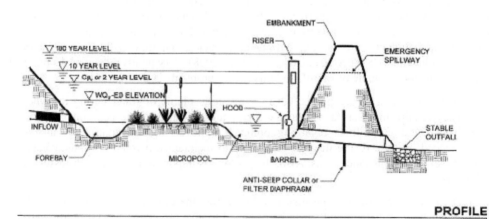

Figure 5.4: Wet pond design (US-EPA)

Table 12: Maintenance activities for dry or wet ponds (source: US-EPA (2011))

Activity	Schedule
• Note erosion of pond banks or bottom	Semiannual inspection
• Inspect for damage to the embankment	
• Monitor for sediment accumulation in the facility and fore-bay	Annual inspection
• Examine to ensure that inlet and outlet devices are free of debris and operational	
• Repair undercut or eroded areas	
• Mow side slopes	Standard maintenance
• Manage pesticide and nutrients	
• Remove litter and debris	
• Seed or sod to restore dead or damaged ground cover	Annual maintenance (as needed)
• Remove sediment from the fore-bay	5- to 7-year maintenance
• Monitor sediment accumulations, and remove sediment when the pond volume has been reduced by 25 percent	25- to 50-year maintenance

5.4.2 Storage Modeling

In order to include the costs related to storage or detention ponds, is necessary to model the storage in the pipe network and its impact on the flooded areas. Most of the existing hydrodynamic models can model storage; they allow for the use of storage nodes where it is possible to define the area of the storage in order to compute the volume for a given value of water depth.

Storage modeling is carried out through a storage node and a throttle in the hydrodynamic model as schematized in Figure 5.5. The storage is defined by its volume which is computed by multiplying the area times the water depth. It is necessary also to include a couple of rules. When the volume in the storage node exceeds the maximum volume assigned to the node, the throttle is closed, and when water depth decreases outside the node and decreases below the maximum storage level the throttle is opened again.

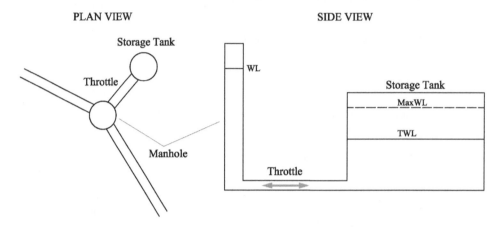

Figure 5.5: Schematization of storage tank simulation

5.4.3   Storage tanks costs

The costs for storage tanks are variable and subject to several factors. The main factor is the material. As mentioned above, in areas with a high density of population it is common that underground structures are built of reinforced concrete. This kind of storage is costly, not only because of the material but also because of the opportunity land cost. Ponds excavated in earth are preferable if possible. The most common way to express the cost of the storage is as a power expression as in Eq 63. This cost includes construction and maintenance costs during the whole life of an asset.

$$C_S = k \cdot V^n \qquad (63)$$

where: C_s is the total tank cost

k is a coefficient

n is an exponent

V is the volume of the tank

US-EPA gives an indication of the costs for wet and dry pond as shown in Eq 59. This costs do not include the land cost. In the case of concrete tanks a price analysis was carried out taking into account materials, equipment and labor; the results are summarized in table 13. A comparison between dry and wet ponds and for concrete tanks is given in Figure 5.6.

$$C_D = 186 \cdot V^{0.70} \quad (Dry \quad ponds)$$
$$C_W = 1680 \cdot V^{0.58} \quad (Wet \quad ponds) \tag{64}$$
$$C_C = 910 \cdot V^{0.88} \quad (Concrete \quad tanks)$$

where: C_D, C_W and C_C is the total tank cost in €

   V is the volume of the tank in m3

Table 13: Cost analysis for concrete storage

Volume (m3)	Cost (€)
75	44,250
340	139,400
1520	562,400
3960	1,386,000
8640	2,764,800

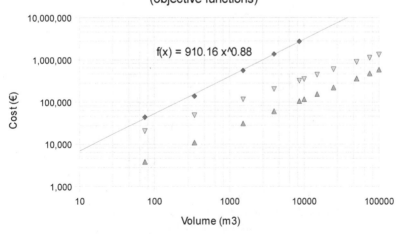

Figure 5.6: Storage costs for concrete, dry and wet ponds

5.5 Diversions in UDS Rehabilitation

Diversions in urban drainage systems are used to divert the water to other places in which it is less harmful. They are commonly used in combined sewer networks, in order to avoid overloading treatment plants. In separated systems they are used to divert water to other points on the network or discharge it directly into the main stream. Usually, diversions are formed as two main structures; a control structure and a conduction structure. Control structures are used to control the water discharge and the level of the water at a specific point of the system. The most common control structures are gates and weirs that can be controlled manually or automatically. Conduction structures convey water to the destination point in the network system; they are composed of pipes or channels.

Diversion structures can be designed on-line or off-line. An on-line design implies that the structure can work under a set of rainfall events with different return periods. An off-line design is optimized to work for a specific return period. In our case an off-line case is assumed. The most simple control structure is an "flow splitter"; this is a manhole designed to split the incoming flow into two parts, usually using a weir structure; see Figure 5.7.

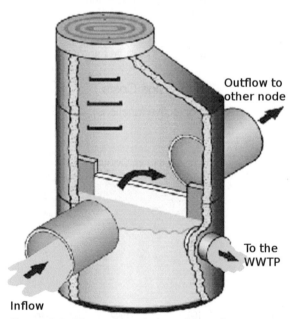

Figure 5.7: Flow splitter (adapted from Storm-water Management Inc)

In order to estimate the dimensions of a diversion the two structures mentioned above have to be modeled using the hydrodynamic model. Most of the hydrodynamic models can do this task. It consists in connecting the intended node with a fictitious node using a weir (orifice or gate), and then connecting this last node with the node that will receive the discharge. The connection is made using a link, which represents the conduction structure.

The cost function for a diversion structure is variable. It depends on many factors; it will depend on whether the structure is for a combined or separated system; it will also depend on whether it is for an online or offline operation. In this thesis it is assumed that the system is for a separate system and with an off-line function; this assumption simplifies the cost function equation. The cost function is divided in two parts: the diversion structure cost and the conduction structure cost. The diversion structure is assumed to be dependent on the depth, and the conduction structure on the diameter of the pipe. As can be seen, depending on the location of the weir, its cost will be a fixed value; however, the conduction structure cost will depend on the diameter of the pipe; in other words, on the discharge to be derived. In order to take this into account the same catalogue used for the pipe renewal is used for the conduction structure and is given by Eq 62. The cost function for the flow splitter and the conduction structure is given by Eq 64.

$$C_d = 1895 \cdot h^{0.9317} + C_{cp} \tag{65}$$

where: C_d is the total diversion cost

                  h is the depth of the split structure

                  C_{cp} is the conduction structure cost (Given by Eq 62)

5.6 Including Other Variables

Until now only the rehabilitation cost has been analyzed in the rehabilitation approach. It is assumed in the study that a multi-objective optimization approach is used, as a consequence other variables have to be included in the analysis. The most obvious variable to include is related to the flooding damages. Damages can be divided into tangible and intangible damages. Tangible damages are those damages that can be evaluated in monetary units while intangible are not easy to evaluate in monetary units; they can be understood better if they are expressed in natural units (Smith and Ward 1998).

$$Damages\begin{cases} Tagible: \begin{cases} \textbf{Direct}: \text{damages to buildings, people belognings, agriculture, etc.} \\ \textbf{Indirect}: \text{emergency costs, traffic disruption, economy loses,etc} \end{cases} \\ Intangible: \begin{cases} \textbf{Direct}: \text{life losses, diseases proliferation, archaeological and cultural damages, etc.,} \\ \textbf{Indirect}: \text{postraumatic and psichological damages, increased vulnerability, etc.,} \end{cases} \end{cases}$$

Figure 5.8: Flooding damages classification

Intangible damages are usually associated with human health, environment, archeological or cultural values. Tangible or intangible damages can be direct or indirect. Direct damages are those produced by the water depth, flooding duration or water velocity effects. On the other hand, indirect damages are the losses produced as a side effect of the flooding (Figure 64). Some authors include primary and secondary damages.

There is a variety of methodologies for flooding damages evaluation. Flood damage estimation can be addressed in four basic steps (Messner et al. 2007):

- Selection of the appropriate approach: this is based on the spatial scale, the objective of the study, data availability, and pre-existing data. These parameters are vital in view of the complexity of the damage function. A detailed damage function requires a high resolution of scales which is costly. A sophisticated damages function also requires a high quality of the data, which may be not available or expensive to measure.

- Determination of the damage categories to be used: these categories are parameters which determine the damage cost function. Tangible and intangible costs have to be included; however, tangible direct costs are those which usually weigh more in the damage function, and more specifically for buildings and inventories. It is a common use to employ simplified function for indirect costs.

- Collection of necessary information: this information is in regard to land use, values at risk, inundation characteristics and the damage function. Land use is related to the property values at risk; depending on the resolution it can be lumped as "industrial", "residential" or "commercial" areas or it can be described in detail items of the characteristics of the buildings in such areas. Also, it is necessary to have a measure of the real value of properties; real value means its remaining value (devalued value) and never the reposition value. Inundation characteristics are related to the frequency and intensity of the flooding; water depth, duration and velocity are the most used variables in the damage estimation. The damage function selection is done depending on the availability of information.

- Calculation of the expected damage: it consists on bringing all the information together and computing the expected flooding damage. At this stage two main approaches can be used: an absolute damage estimation and the percentage of property value; first compute the value in monetary cost as an amount for each property; the second approach is to compute the relative value of the property related to a maximum expected value.

No matter which approach is used, the function damage cost for flooding can be expressed as in Eq 60.

$$Damage_{total} = \sum_{i=1}^{n} \sum_{j=1}^{m} \left(value_{i,j} \times susceptibility_{i,j} \right)$$

$$(66)$$

$$susceptibility_{i,j} = f \left(Entity_{i,j}, Inundation_{k}, Socioeconomic_{l} \right)$$

where: $Damage_{total}$ is the total flooding damage

i	is the category of the element at risk
j	is the element at risk
k	is the flood type or scenario
l	is type of socioeconomic system
$value_{i,j}$	is the value of the element at risk
$Susceptibility_{i,j}$	is the percentage of damage parameter
$Entity_{i,j}$	is the characteristic of the element at risk
$Inundation_k$	is the characteristic of the inundation
$Socioeconomic_l$	is the characteristic of the socioeconomic

In this thesis several types of damage equation are explored taking into account that they are more oriented to be used in developing countries. Developing countries are characterized by the lack of data and funds to collect it. In the following sections several damage functions are used ranging from direct tangible cost, quality evaluation to intangible evaluation through some test examples.

5.7 Pipe Renewal vs Damage Costs

This section is aimed at developing and comparing a multi-objective analysis using several methodologies to rehabilitate a pipe network. The problem presented here is posed as a rehabilitation optimization problem for an existing UDS, where the minimization of the pipe rehabilitation costs (below-ground system) and the minimization of the surcharge-related damage costs (above-ground system) are considered. For simplicity reasons only these two objectives are selected. The same approach can be applied to a more complex system having a larger number of objectives.

The network system contains 12 pipes, 13 manholes and 11 sub-catchments. A raster GIS map is used representing the land use in order to establish the unit cost as a function of the building and structure types. Using a topographic map, sub-catchments are delineated and assigned to the nearest node in the network system. Figure 5.9 shows the schematization of the network and the layers representing land use and the catchments. Table 14 and 15 give details of the catchments and the pipe network.

In addition to the delineation of the catchments, curves of flood depth, area and volume are constructed for each catchment. This is done by combining a raster GIS tool with an additional software tool implemented in Borland Delphi to compute costs and build the nodal depth-area-volume-cost tables. The system is a storm-water system, so the dry weather flow due to waste-water is not considered. The rainfall used corresponds to an event of 2.77 mm depth over 6 hours duration.

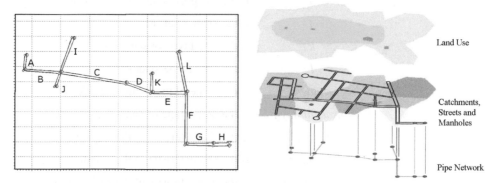

Figure 5.9: Network topology, land use, and above - below ground network system

Table 14: Catchment description

Node Id	Area (has)	Inhabitants
0014_1	4.43	1116
0030_1	3.71	935
0078_1	1.76	444
K001_1	1.44	363
K002_1	0.35	88
K003_1	2.65	668
K006_1	1.57	396
0051_1	3.61	910
0073_1	5.08	1280
0127_1	6.39	1610
0140_1	4.74	1194

Table 15: Pipe network description

Link Id	Diameter (m)	Length (m)	Slope $^0/_{00}$	Selected
A	0.225	15.81	3.79	no
B	0.225	173.31	0.92	no
I	0.225	126.87	1.26	no
J	0.225	173.31	0.58	no
C	0.225	441.62	0.20	yes

Link Id	Diameter (m)	Length (m)	Slope ‰	Selected
D	0.225	491.17	0.20	yes
K	0.225	610.63	0.13	no
E	0.225	616.55	0.18	yes
L	0.225	751.81	0.21	no
F	0.225	788.00	0.11	yes
G	0.225	881.61	0.07	yes
H	0.450	930.51	0.03	yes
As	Street	15.81	6.32	---
Bs	Street	173.31	-0.58	---
Cs	Street	610.63	0.34	---
Ds	Street	457.00	0.00	---
Es	Street	616.55	-0.03	---
Fs	Street	788.00	-0.06	---
Gs	Street	881.61	0.16	---
Is	Street	120.21	-0.17	---
Js	Street	173.31	-0.58	---
Ks	Street	610.63	0.34	---
Ls	Street	751.81	0.25	---
Hs	Street	940.90	0.29	---

5.7.1 Hydraulic simulation model

The computation of the objective function's values is based on the hydrodynamic simulations of the water motion in the pipe network. For this purpose the MOUSE software package (developed by DHI Water and Environment) is used. This modeling system solves the Saint Venant equations (1D); it was selected due its robustness and the possibility of connecting it with the two-dimensional (2D) modeling system MIKE21 for the surface water simulation.

In addition to the below-ground pipe network system, the above-ground or major system, representing the streets, is simulated with the same 1D model. If the water level at a node reaches the ground level, water flows onto the street over a weir with the crest level at the ground level and a crest length equal to the manhole perimeter. Water can flow along the streets and back into the pipe network system or flood other areas when the water level in the street is above the curb level. The above-ground model is complemented by a simplified model for water levels greater than the curbs. In this model excess water at a node fills an inverted cone with a prescribed level-area-volume-cost table constructed for this purpose at each node in the street network.

5.7.2 Choice of objective functions

Two objective functions are evaluated: rehabilitation investment cost and damage cost due to flooding. Investment cost and flood damage can be measured in monetary terms, but it should be noted that different stakeholders carry these costs. These two objectives are conflicting: more investment reduces damages and vice-versa. An interface software in the platform was developed in order to compute the objectives (see Chapter 3). The interface runs the hydrodynamic model, selects the critical pipes for the renewal, computes investment cost and calculates flood damage costs. Finally, it returns control to the optimizer.

The problem to be solved is deliberately made relatively simple: the size of the network is small (12 pipes and 12 streets links); only pipe renewal is considered as a rehabilitation technique; all pipes are costed in the same way, independent of their spatial location; the expenditure is not phased in time and only one design rainfall is considered. These assumptions are made in order not to complicate the presented methodology with too many details.

5.7.3 Investment Cost

In this example, which should be viewed as providing a "proof of concept", we use simplified formulations of the objective functions. For rehabilitation, only pipe replacement is considered. A table of replacement costs for commercial pipe diameters per unit length is used to calculate the overall costs. These unit costs include: pipe cost, excavation, installation of the pipe per unit length and restoration and reinstatement.

Such costs are normalized using the maximum cost expected for the rehabilitation. The maximum cost is computed assuming that it is necessary to change all the pipes that are allowed to change for the largest diameter from the catalog of pipes. The mathematical expression of the function (f_1) for the underground costs can be written as:

$$f_1 = \frac{\sum_{i=1}^{n} C_i \cdot L_i}{\sum_{i=1}^{n} C_i^{max} \cdot L_i} \tag{67}$$

where: L is the pipe length, C is the pipe replacement cost and C^{max} is the maximum pipe cost.

5.7.4 Flood Damages

The flood damage cost is computed using the following procedure. For each node the maximum spilled volume is determined from the hydraulic model. Using this volume and the pre-prepared table of level-area-volume-cost for each sub-catchment, the level is computed by interpolation. The depth for a single cell is computed by subtracting this level from the ground level given by the topographic raster map.

In order to evaluate the risks to communities, properties and infrastructure effectively, it is important to estimate the distribution of hazards and the magnitudes of flood-related damages. Generally, such damages are divided into tangible and intangible damages. Those damages that can be estimated and expressed directly in monetary terms are the so-called tangible damages (e.g., damages to properties, infrastructure, etc.). Within the scope of this study case only this type of damages is considered. There are several variables that can be used for analyzing the tangible damages, the most common are: depth, velocity and duration of flooding. Also, land use is of major importance as it defines the type, cost and vulnerability of structures. There are several equations that can be used to estimate flood damage costs in urban areas (Tang et al 1992). In this thesis we use the following relationship:

$$FLD = a + b \cdot DEP + c \cdot DUR \tag{68}$$

where: FLD is the flooding damages per unit area, a, b and c are coefficients related to land use, DEP is the water depth, and DUR is the flooding duration. Flooding duration is also neglected and the value of a is assumed to be zero.

However, the value of b is variable: $b = fc \cdot C_{max}$ where fc is a factor dependent on the depth (DEP), and C_{max} is the maximum cost damage per unit area. The values are taken from Van der Sande et al (2003) and correspond to studies carried out in The Netherlands on flood damage around the Meuse River. The associated graphs were digitized and located in a file for use with the model integration.

Finally, the equation for flood damages can be expressed as:

$$f_2 = \frac{\sum_{i=1}^{ncells} \sum_{k=1}^{nlu} (fc_i^k \cdot C_{max}^k \cdot DEP_i)}{\sum_{i=1}^{ncells} \sum_{k=1}^{nlu} (fc_i^k \cdot C_{max}^k \cdot DEP_{max})} \tag{69}$$

where: f_2 is the total flood damage, $ncells$ is the total number of cells in the raster map, nlu is the total number of land use types, fc_i^k is the value of damage factor at i^{th} cell with k^{th} land use type, and DEP_i is the depth at i^{th} cell.

Similarly, if a third objective is introduced then the new objective function needs to be added and formulated to penalize unacceptable values. For example, when dealing with urban drainage and sewerage systems another objective in addition to flood damage and infrastructure costs could be to minimize the pollution from sewer overflows. This objective could be added as another function and those results for which the concentration levels coming from sewer overflows exceed acceptable environmental standards would be penalized. Also intangible damages can be included such as proposed by Lekuthai and Vongvisessomjai (2001). See Ellis et al (2004) for a list of objectives that can be included using this methodology.

5.7.5 Results and Discussion

The first automatic diagnostic run shows that six (6) pipes are suitable for replacement. They are selected on the basis of the relationship between the actual discharge and the capacity of the pipe (Q/Qf). This ratio is directly related to the surcharge in the pipes and the possible bottlenecks in the system. The practitioner can of course change these selected pipes directly on the basis of his/her own judgment and expertise. For this case the relation Q/Qf was set in 2; pipes carrying twice its discharge capacity are candidates for replacement.

The optimization process using several algorithms was carried out using a combination of population size and generations to give around 1200 function evaluations. From Table 16, it can be seen that the NSGA-II algorithm has a better cardinality than ε-MOEA, increasing with population size, while ε-MOEA has a more uniform distribution of points in the Pareto with the cardinality remaining approximately the same, no matter what the population size. The computational time for both algorithms increases considerably with the increase in population size. Table 17 shows that when compared with ε-MOEA the hypervolume for NSGA-II is similar, however, in terms of dispersion ε-MOEA is worse than NSGA-II, giving more variability in the results.

Table 16: Metrics indicators for NSGA-II and ε-MOEA, cardinality and execution time

Method	Population Size	Cardinality		Time (min)	
		Average	Standard deviation	Average	Standard deviation
NSGA II	20	13.4	1.43	143.8	6.3
	32	23.2	1.99	186.1	8.5
	64	42.3	4.60	284.3	10.9
ε-MOEA	20	17.2	2.94	151.4	5.9
	32	19.2	1.69	239.1	11.6
	64	18.7	3.53	290.1	14.8

Table 17: Metrics indicators for NSGA-II and ε-MOEA, hypervolume and ε-indicator

Method	Population Size	Hypervolume		ε-indicator	
		Average	Standard deviation	Average	Standard deviation
NSGA II	20	1.3613	0.0069	1.0124	0.0068
	32	1.3689	0.0031	1.0101	0.0069
	64	1.3713	0.0048	1.0052	0.0035
ε-MOEA	20	1.3426	0.0189	1.0031	0.0040
	32	1.3590	0.0073	1.0037	0.0050
	64	1.3559	0.0137	1.0050	0.0064

When comparing the Pareto sets produced by successive generations, the metric indicators, hyper-volume and ε-indicator, show convergence. This feature makes it possible to use them as a stopping criterion. However, the ε-indicator has more variability between successive generations which makes it difficult to be used as a stopping criterion, whereas the hyper-volume presents a smoother convergence for successive generation (Figures 3 and 4). NSGA-II shows larger values of hyper-volume than ε-MOEA, indicating that the Pareto set generated by NSGA-II covers a larger space than ε-MOEA. The ε-indicator shows better values for ε-MOEA.

Another interesting result is that it was possible to find a wide diversity of solutions, including the "*do nothing*" solution (100% damages) found by NSGA-II, which was not found by ε-MOEA. However, the method was capable of finding flooding damage values under 90% of the maximum damage. Objectives in Figure 5.10 and 5.11 correspond to the normalized rehabilitation cost and the flood damages. If there is no investment in rehabilitation, the flood damage will be a maximum and equal to 1.00. On the other hand, if the investment is around *0.22* of the maximum theoretical costs, no flood damage will be expected. In between, there are a large number of other Pareto-optimal solutions. This gives decision makers a wide range of options and in the case of constrained resources for rehabilitation the possibility to select a trade-off region.

From Figure 5.19 and Table 18 it can be seen that it is possible to compute a set of equally valuable solutions (Pareto set) after removing illogical or undesired solutions through post processing. Initially a penalty function was introduced to the objective functions at optimization time; however, it affected the performance and convergence of the MOEA process considerably. For this reason it was decided to develop a post-processing tool instead that selects a limited number of solutions from the Pareto set for further analysis. Visualizations, as presented in Figure 5.19, allow users, stakeholders and decision makers to define a base of optimal solutions for negotiation and to set the maximum limit of acceptance for each actor. For instance, an investor may want to reduce the cost of the pipes, whereas the householders seek to diminish the flooding damages; they can reach an agreement using the set of optimal solutions inside their ranges of acceptance.

More information can be extracted from Figure 5.19. The slope of the Pareto front becomes less steep at the point (0.20; 1.00); this indicates a reduction of the gain per invested unit. For example, to considerably reduce the flood damage from \$7.1 Millions to \$1.0 Millions an investment of \$0.2 Million is needed. For larger investment costs, in order to reduce the remaining possible flood damages (about 10%) the same amount (or more) has to be invested.

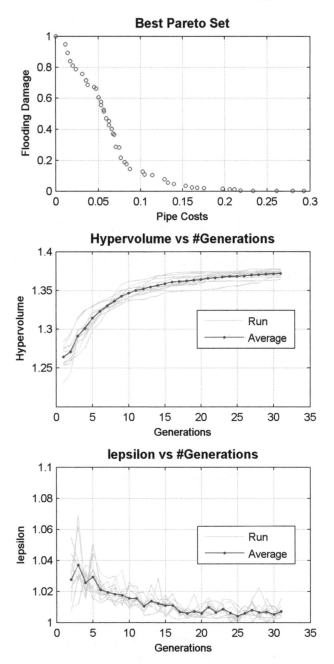

Figure 5.10: Pareto front and metrics for NSGA-II

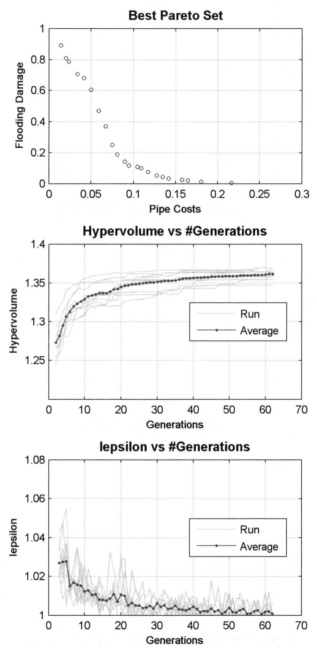

Figure 5.11: Pareto front and metrics for ε-MOEA

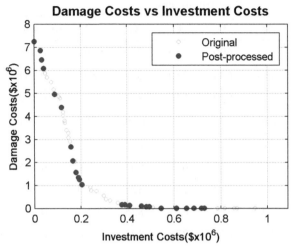

Figure 5.12: Pareto front after post-processing

Table 18: Post-processed solutions

Solution	Pipe Diameters						Costs	
	C	D	E	F	G	H	Investment	Damages
1	225	225	225	225	225	450	0.000	7.240
2	225	225	225	225	300	450	0.027	6.861
3	225	225	225	225	375	450	0.033	6.451
4	225	225	225	225	450	450	0.040	6.073
5	225	225	225	225	600	600	0.088	4.957
6	225	225	225	450	450	450	0.117	4.385
7	225	225	225	450	525	600	0.157	2.672
8	225	225	225	450	600	600	0.165	2.065
9	225	225	225	525	600	600	0.179	1.567
10	225	225	225	525	600	750	0.189	1.345
11	225	225	225	600	600	600	0.194	1.255
12	225	225	225	600	600	750	0.204	1.039

Solution	Pipe Diameters						Costs	
	C	D	E	F	G	H	Investment	Damages
13	225	525	600	750	750	750	0.376	0.167
14	225	450	600	750	825	900	0.391	0.164
15	225	525	600	750	825	1050	0.410	0.135
16	225	600	600	750	1050	1200	0.461	0.107
17	225	600	825	825	825	1200	0.482	0.083
18	225	600	825	825	900	1200	0.495	0.079
19	450	525	750	750	825	1200	0.545	0.011
20	600	750	825	825	825	825	0.610	0.000
21	750	750	825	825	825	825	0.654	0.000
22	750	750	825	825	825	1200	0.685	0.000
23	750	750	825	825	1050	1200	0.716	0.000
24	750	750	750	900	1050	1200	0.728	0.000

5.7.6 Test case using 63 pipes

This case was set up using the same network configuration described in the previous case with twelve pipes but increasing its number to 63 pipes. The main objective is to test the tool and procedures for a more complex network system. The network system this time is formed by 63 pipes, 65 manholes and 8 sub-catchments, the network corresponds to a simplified drainage network located in Denmark. This time only the NSGA-II algorithm was used in the optimization procedure due to its better global performance in the previous case with 12 pipes. The total number of function evaluations were 1920; population size was set to 24 and number of generation 80. The rest of the parameters remain the same as for the 12 pipes case.

Results and Analysis

The first step was to run the tool for a diagnostic of the system. It was able to identify four pipes to be changed, taking into account the relation between the theoretical capacity of a pipe under steady state conditions and its actual maximum discharge (Q/Qf). Figure 5.13 shows the sewer network and the location of selected pipes for replacement. The ratio Q/Qf was set into 1.2 in this case. Flooding can be seen at three nodes (Figure 5.13). Two of the nodes belong to the selected pipes for replacement and the third one is located in a pipe upstream that was not selected. A fourth pipe was also automatically identified and selected to be changed, but it does not have any flooding associated with it; this occurs due to the ground level being high and there is no flooding, however the pipe works as a pressurized pipe.

From the profile shown in Figure 5.14 it can be seen how the first selected pipe, from down-stream to upstream, is under-designed, acting as a bottleneck. The high gradient of the energy is an indication that such pipes are working under extreme conditions and are required to be upgraded.

For this particular case, the Pareto front shows a singular convex shape; see the Figure 5.15. If an imaginary trend line is drawn in this figure, the line will be a convex line, which is one of the special cases in the optimization referred to in Chapter 4. Also, it can be identified with the three main groups of solutions in Figure 5.15. One, ranging from 0.00 to 0.012 of the investment cost, is called range one (R-1), an other, ranging from 0.037 to 0.046, is range two (R-2), and range three (R-3) ranges from 0.05 to 0.06 of the investment cost. These ranges are enclosed in a dashed line in Figure 5.15.

From the analysis of the Pareto set and the decision space, it can be seen that each range or group of solutions belongs to a pipe to be changed. For instance, analyzing the decision space it can be deduced that the pipe "P-1" generates all the solutions enclosed in the range R-1; this pipe dominates the investment cost in the whole range R-3. Pipe "P-1" has a very small length and its influence on investment cost is low; the diameter can be changed without a high increase in investment costs and is preferred to be changed rather than other more expensive pipes. For this reason the groups or clusters are formed in the Pareto. The lowest investment is obtained by changing pipe "P-1"; at the beginning it improves the damage cost at a high rate, with a very steep slope from damage cost 1.0 to 0.7; but at the end of the range "R-1" the slope becomes milder and almost horizontal. This means that with more investment fewer reduction in damage cost is obtained, and no matter how big the pipe diameter is, there is no major improvement in the damage due to flooding. When pipe "P-2" is changed (increased), the solution jumps from range "R-1" to "R-2"; this diameter is combined with several diameters of pipe "P-1" in the decision space; once again pipe P-1 dominates the rest of the solutions using other diameters. However, R-2 retains a more constant slope where the investment for the rehabilitation is almost linear due to the improvement in the damage cost. The same behavior is observed for cluster "R-3", but this time there are some solutions which are very close to zero damage, showing a weak dominance; in this case pipes "P-1" and "P-2" again dominate the other solutions.

The optimization process shows a logical behavior. It performs as well as a practitioner could do under the same circumstances, identifying the problematic pipes, that are not necessarily located at the nodes that are flooded. Also, it is able to test a set of solutions that improve the damage cost by optimizing the investment cost. Figures 5.16 to 5.18 show how the flooding is reduced by the change in diameters of pipes "P-1" and "P-2"; the other two additional pipes, identified as being problematic ("P-3 and pipes with no flooding in the nodes) remain without change because they have no influence on the flooding damage costs. Regarding the execution time, it has increased by about 237% percent when compared with the twelve pipes case. This is not only due to the increasing in the number of function evaluations from 1200 to 1900 but also to the increase in the total number of pipes and nodes to be included in the hydraulic model. The total run time was on average 9.7 hours for a single run despite the number of diameters to be changed, which they were less than the twelve pipes study case. It is expected that the more pipes to optimize, then the more computation time will be required.

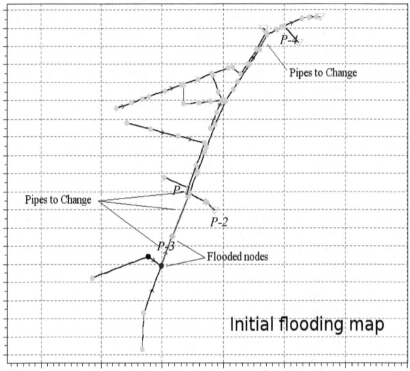

Figure 5.13: Pipes to change and flooded nodes

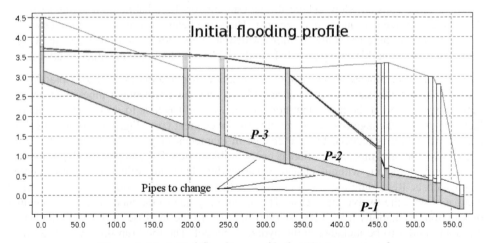

Figure 5.14: Initial flooding profile for 63 pipes network

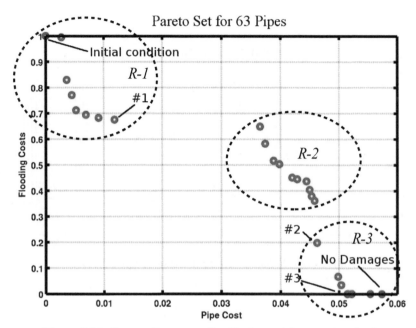

Figure 5.15: Convex Pareto set for 63 pipes with normalized values

Table 19: Diameters for initial and optimal solutions over the Pareto

Case	Diameters (mm)		
	P-1	**P-2**	**P-3**
Initial Conditions	300	300	300
Solution #1	600	300	300
Solution #2	350	350	300
Solution #3	450	375	300
No damages	400	400	300

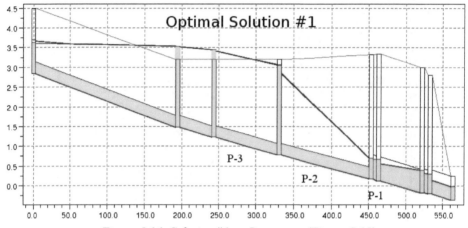

Figure 5.16: Solution #1 on Pareto set (Figure 5.15)

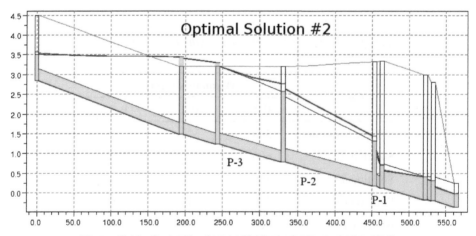

Figure 5.17: Optimal solution #2, increasing pipe p1 and p2

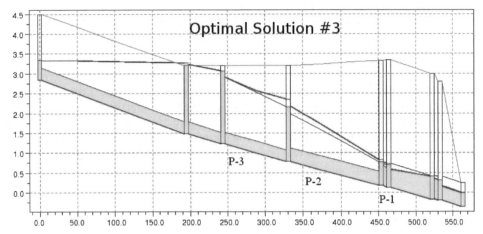

Figure 5.18: Optimal solution #3, increasing pipe P-1 respect solution #2

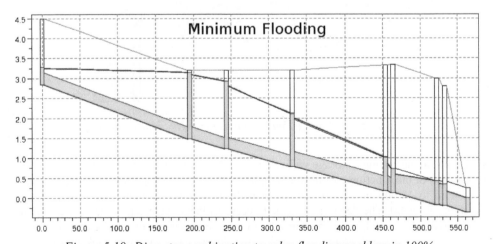

Figure 5.19: Diameter combination to solve flooding problem in 100%

5.8 Use of tank storage as rehabilitation measure

As mentioned above one of the most common measures that can be used in the rehabilitation of urban drainage networks is the use of storage tanks, and wet or dry ponds. The incorporation of storage tanks or wet and dry ponds into the developed framework was done using a storage node and a throttle in the model as it is explained in Section 5.4.2 and cost functions from Section 5.4.3 .

This approach was tested on the twelve pipes example, this time adding six possible locations for wet pond or concrete tank storage as is shown in Figure Error: Reference source not found. The storage locations have to be settled by a practitioner with expertise. As shown, the tanks were located on the backbone of the network in order to store enough water and to reduce the diameter this way. The optimizer of the developed multi-objective framework has to decide between two options: in the first place it has to decide whether or not to activate the storage, and second it has to find the optimal dimensions of the tanks and throttle.

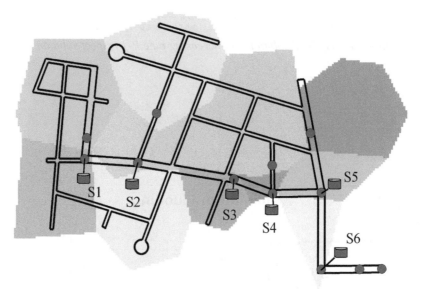

Figure 5.20: Storage tank locations for twelve pipe example

In order to avoid having a storage with a very small volume, which can be inconvenient due to the form of the function cost power equation, a minimum storage was set to 85 m^3, and a maximum to 10,000 m^3. These values keep the results in the range for which the cost functions are valid. The depth-area-capacity curve of the storage can be defined depending on the hydraulic model. Some models allow for the introduction of tabular data and others only require a prismatic volume given the area. In our case a circular tank was assumed with the necessary depth to store the maximum volume. A throttle controls the inflow or outflow of water; if the level of the storage is lower than the connected manhole, and there is still capacity in the storage tank, then water will enter into the tank; if the water level in the manhole is lower than in the storage tank the throttle will be open, allowing the water go out of the tanks. The developed algorithm is able to select the dimensions of the throttle.

The same NSGA-II parameters were used for the multi-objective optimization as the case presented in Section 5.7 . The rehabilitation cost is composed of two values and is the result of adding pipe renewal cost and storage cost. The damage cost function remains the same. The optimization procedure was run 10 times, using a population size of 100 during 100 generations, producing 10000 function evaluations. This was more than those used for the case without storage tanks.

Results and Analysis

A single run of the optimization process took about 27 hours in average, using a dual core 2.4GHz PC. As can be argued the problem escalates with the increase in population size in the GA parameters and the increase of the variables to be optimized. Now there are six variables more if storage tanks are included. Two different setups of the model were used based on land use. When there is the possibility to use the land, dry or wet ponds may be used, but when the area is highly urbanized, concrete tanks are preferred due to the opportunity cost for the land.

The first setup was using wet ponds and pipe renewal as measure for drainage network rehabilitation, and the second was the use of concrete tanks and also pipe renewal as a rehabilitation technique. Figure 5.17 shows the Pareto front when wet ponds are included in the optimization process. It can been seen how in this Pareto optimal solutions outperform the Pareto optimal solution when only pipe renewal is used. Also, there is a very well spread Pareto set when pipes and wet ponds are used, showing a higher cardinality and diversity than the Pareto set when only pipe renewal is used. However, when post-processed solutions are analyzed in more detail (Table 18),   it can be seen that only storages were used and no pipe has changed its original value. This behaviour of the optimization results is reasonable; a wet pond is a cheaper solution than the renewal of the pipes.

In order to check the behavior of the Pareto set, the optimization with concrete tanks was used instead of wet ponds, and a second experiment was conducted. The cost function was computed by adding pipe renewal and the cost function equation for concrete tanks as shown in Eq. 59. This time the Pareto set, when pipe renewal and concrete tanks is used, did not outperform fully the Pareto set for only pipe renewal as the rehabilitation technique. Storage tanks are expensive and their use is costly compared with renewal of pipes. From Figure Error: Reference source not found, it is seen that for low investment, storage tanks are used and dominate the only pipe renewal Pareto set, but after investment reaches € 150,000, the use of pipe renewal outperforms the solutions with the storage tanks included.

It was expected that the Pareto set would follow an envelope line as shown in Figure Error: Reference source not found (dashed line), which overlaps both Pareto sets. However, despite the number of runs; it was not possible to converge to the expected Pareto without redirecting the optimization with additional constraints or penalty functions. On the other hand, when the results are analyzed in detail (Table 19), it can be seen that the values are not far from the envelope line, and still the results provide a good information for decision-makers and stakeholders. In the detailed results a variety of solution can be distinguished: solutions without any change in the pipes, solutions not using storage tanks, and combined solutions with pipe changes and storage tanks; their expresses the diversity in the solutions.

When only pipe renewal was used as a rehabilitation technique 24 solutions were found, while for storage and pipe renewal 67 and 70 are found for wet ponds and concrete tanks respectively. These give a wider range of solutions to decision-makers. Wet ponds range from a volume of 85 to 5400 m³ and concrete storage tanks range from 85 to 650 m³. If the corresponding prices of the storage are computed, it can be seen that both values are about 250,000 € (245500 and 268200 respectively). This indicates that there is a fixed upper limit to the investment in storage tanks. If this upper limit is exceeded it is better to use pipe renewal as the rehabilitation measure.

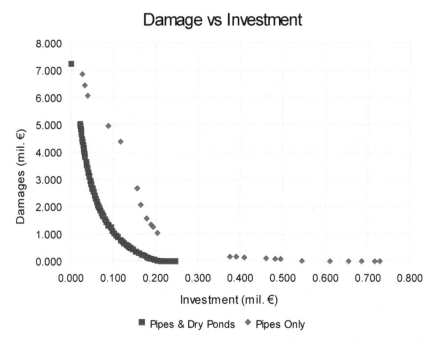

Figure 5.21: Comparison between the use of wet ponds & pipes renewal against the use of only pipe renewal as rehabilitation technique

Table 20: Optimization results for dry ponds and pipe renewal as rehabilitation technique

	Pipe Diameter (mm)						Storage (m³)						Cost (€x10⁶)	
Sol. #	C	D	E	F	G	H	S1	S2	S3	S4	S5	S6	Investment	Damages
1	225	225	225	225	225	450	0	0	0	0	0	0	0.000	7.240
2	225	225	225	225	225	450	85	0	0	0	0	0	0.022	5.025

	Pipe Diameter (mm)						Storage (m³)						Cost (€x10⁶)	
Sol. #	C	D	E	F	G	H	S1	S2	S3	S4	S5	S6	Investment	Damages
3	225	225	225	225	225	450	89	0	0	0	0	0	0.023	4.962
4	225	225	225	225	225	450	96	0	0	0	0	0	0.024	4.848
5	225	225	225	225	225	450	101	0	0	0	0	0	0.025	4.767
6	225	225	225	225	225	450	108	0	0	0	0	0	0.025	4.673
7	225	225	225	225	225	450	122	0	0	0	0	0	0.027	4.472
8	225	225	225	225	225	450	130	0	0	0	0	0	0.028	4.378
9	225	225	225	225	225	450	141	0	0	0	0	0	0.030	4.243
10	225	225	225	225	225	450	150	0	0	0	0	0	0.031	4.144
11	225	225	225	225	225	450	153	0	0	0	0	0	0.031	4.103
12	225	225	225	225	225	450	162	0	0	0	0	0	0.032	4.015
13	225	225	225	225	225	450	171	0	0	0	0	0	0.033	3.925
14	225	225	225	225	225	450	182	0	0	0	0	0	0.034	3.823
15	225	225	225	225	225	450	202	0	0	0	0	0	0.037	3.641
16	225	225	225	225	225	450	216	0	0	0	0	0	0.038	3.528
17	225	225	225	225	225	450	232	0	0	0	0	0	0.040	3.409
18	225	225	225	225	225	450	255	0	0	0	0	0	0.042	3.242
19	225	225	225	225	225	450	273	0	0	0	0	0	0.044	3.138
20	225	225	225	225	225	450	277	0	0	0	0	0	0.044	3.112
21	225	225	225	225	225	450	306	0	0	0	0	0	0.047	2.945
22	225	225	225	225	225	450	325	0	0	0	0	0	0.048	2.849
23	225	225	225	225	225	450	333	0	0	0	0	0	0.049	2.814
24	225	225	225	225	225	450	362	0	0	0	0	0	0.051	2.675
25	225	225	225	225	225	450	367	0	0	0	0	0	0.052	2.651
26	225	225	225	225	225	450	393	0	0	0	0	0	0.054	2.548
27	225	225	225	225	225	450	431	0	0	0	0	0	0.057	2.407
28	225	225	225	225	225	450	464	0	0	0	0	0	0.059	2.291
29	225	225	225	225	225	450	497	0	0	0	0	0	0.062	2.180
30	225	225	225	225	225	450	536	0	0	0	0	0	0.064	2.063
31	225	225	225	225	225	450	563	0	0	0	0	0	0.066	1.994
32	225	225	225	225	225	450	585	0	0	0	0	0	0.068	1.939
33	225	225	225	225	225	450	617	0	0	0	0	0	0.070	1.858
34	225	225	225	225	225	450	658	0	0	0	0	0	0.073	1.768
35	225	225	225	225	225	450	689	0	0	0	0	0	0.075	1.702
36	225	225	225	225	225	450	726	0	0	0	0	0	0.077	1.635
37	225	225	225	225	225	450	789	0	0	0	0	0	0.081	1.529
38	225	225	225	225	225	450	847	0	0	0	0	0	0.084	1.438
39	225	225	225	225	225	450	915	0	0	0	0	0	0.088	1.339
40	225	225	225	225	225	450	942	0	0	0	0	0	0.089	1.299
41	225	225	225	225	225	450	1054	0	0	0	0	0	0.095	1.238
42	225	225	225	225	225	450	1110	0	0	0	0	0	0.098	1.104
43	225	225	225	225	225	450	1179	0	S3	0	S5	0	0.102	1.026
44	225	225	225	225	225	450	1275	0	0	0	0	0	0.107	0.942
45	225	225	225	225	225	450	1358	0	0	0	0	0	0.111	0.888
46	225	225	225	225	225	450	1512	0	0	0	0	0	0.118	0.771
47	225	225	225	225	225	450	1613	0	0	0	0	0	0.122	0.721
48	225	225	225	225	225	450	1713	0	0	0	0	0	0.126	0.658
49	225	225	225	225	225	450	1838	0	0	0	0	0	0.132	0.596

Sol. #	Pipe Diameter (mm)						Storage (m³)						Cost (€x10⁶)	
	C	D	E	F	G	H	S1	S2	S3	S4	S5	S6	Investment	Damages
50	225	225	225	225	225	450	1931	0	0	0	0	0	0.136	0.567
51	225	225	225	225	225	450	2003	0	0	0	0	0	0.138	0.518
52	225	225	225	225	225	450	2080	0	0	0	0	0	0.142	0.485
53	225	225	225	225	225	450	2221	0	0	0	0	0	0.147	0.439
54	225	225	225	225	225	450	2408	0	0	0	0	0	0.154	0.355
55	225	225	225	225	225	450	2599	0	0	0	0	0	0.161	0.296
56	225	225	225	225	225	450	2802	0	0	0	0	0	0.168	0.236
57	225	225	225	225	225	450	2961	0	0	0	0	0	0.174	0.204
58	225	225	225	225	225	450	3140	0	0	0	0	0	0.180	0.146
59	225	225	225	225	225	450	3250	0	0	0	0	0	0.183	0.118
60	225	225	225	225	225	450	3413	0	0	0	0	0	0.189	0.088
61	225	225	225	225	225	450	3686	0	0	0	0	0	0.197	0.052
62	225	225	225	225	225	450	3912	0	0	0	0	0	0.204	0.017
63	225	225	225	225	225	450	4225	0	0	0	0	0	0.213	0.002
64	225	225	225	225	225	450	4537	0	0	0	0	0	0.222	0.001
65	225	225	225	225	225	450	4923	0	0	0	0	0	0.233	0.001
66	225	225	225	225	225	450	5008	0	0	0	0	0	0.236	0.001
67	225	225	225	225	225	450	5446	0	0	0	0	0	0.247	0.001

Table 21: Optimization results for dry ponds and pipe renewal as rehabilitation technique

Sol#	Pipe Diameter (mm)						Storage (m³)						Cost (€x10⁶)	
	C	D	E	F	G	H	S1	S2	S3	S4	S5	S6	Investment	Damages
1	225	225	225	225	225	450	0	0	0	0	0	0	0.000	7.240
2	225	225	225	225	300	450	0	0	0	0	0	0	0.042	6.871
3	225	225	225	225	225	450	85	0	0	0	0	0	0.048	5.022
4	225	225	225	225	225	450	90	0	0	0	0	0	0.050	4.952
5	225	225	225	225	225	450	95	0	0	0	0	0	0.053	4.872
6	225	225	225	225	225	450	104	0	0	0	0	0	0.057	4.729
7	225	225	225	225	225	450	109	0	0	0	0	0	0.060	4.649
8	225	225	225	225	225	450	116	0	0	0	0	0	0.062	4.560
9	225	225	225	225	225	450	126	0	0	0	0	0	0.067	4.420
10	225	225	225	225	225	450	133	0	0	0	0	0	0.070	4.335
11	225	225	225	225	225	450	140	0	0	0	0	0	0.073	4.256
12	225	225	225	225	225	450	142	0	0	0	0	0	0.074	4.229
13	225	225	225	225	225	450	151	0	0	0	0	0	0.078	4.141
14	225	225	225	225	225	450	159	0	0	0	0	0	0.082	4.050
15	225	225	225	225	225	450	168	0	0	0	0	0	0.086	3.955
16	225	225	225	225	225	450	173	0	0	0	0	0	0.088	3.906
17	225	225	225	225	225	450	183	0	0	0	0	0	0.092	3.810

	Pipe Diameter (mm)						Storage (m³)						Cost (€x10⁶)	
Sol#	C	D	E	F	G	H	S1	S2	S3	S4	S5	S6	Investment	Damages
18	225	225	225	225	225	450	188	0	0	0	0	0	0.094	3.767
19	225	225	225	225	225	450	199	0	0	0	0	0	0.099	3.670
20	225	225	225	225	225	450	206	0	0	0	0	0	0.102	3.611
21	225	225	225	225	225	450	219	0	0	0	0	0	0.107	3.511
22	225	225	225	225	225	450	229	0	0	0	0	0	0.112	3.431
23	225	225	225	225	225	450	238	0	0	0	0	0	0.115	3.365
24	225	225	225	225	225	450	253	0	0	0	0	0	0.121	3.259
25	225	225	225	225	225	450	268	0	0	0	0	0	0.127	3.177
26	225	225	225	225	225	450	280	0	0	0	0	0	0.132	3.096
27	225	225	225	225	225	450	294	0	0	0	0	0	0.138	3.010
28	225	225	225	225	225	450	310	0	0	0	0	0	0.144	2.944
29	225	225	225	225	225	450	314	0	0	0	0	0	0.146	2.912
30	225	225	225	225	225	450	362	0	0	0	0	0	0.165	2.689
31	225	225	225	225	225	450	373	0	0	0	0	0	0.169	2.639
32	225	225	225	225	225	450	398	0	0	0	0	0	0.178	2.582
33	225	225	225	225	225	450	417	0	0	0	0	0	0.185	2.461
34	225	225	225	225	225	450	432	0	0	0	0	0	0.191	2.412
35	225	225	225	225	225	450	449	0	0	0	0	0	0.198	2.343
36	225	225	225	225	225	450	468	0	0	0	0	0	0.205	2.276
37	225	225	225	225	225	450	495	0	0	0	0	0	0.215	2.185
38	225	225	225	225	225	450	521	0	0	0	0	0	0.224	2.112
39	225	225	225	225	225	450	537	0	0	0	0	0	0.230	2.059
40	225	225	225	225	225	450	553	0	0	0	0	0	0.236	2.015
41	225	225	225	225	375	450	462	0	0	0	0	0	0.254	1.994
42	225	225	225	225	375	450	470	0	0	0	0	0	0.257	1.970
43	225	225	225	225	225	450	617	0	0	0	0	0	0.259	1.880
44	225	225	225	225	225	450	648	0	0	0	0	0	0.270	1.795
45	225	225	225	525	600	600	0	0	0	0	0	0	0.277	1.577
46	225	225	225	525	600	600	0	0	0	0	0	0	0.277	1.577
47	225	225	225	525	600	750	0	0	0	0	0	0	0.293	1.354
48	225	225	225	525	600	750	0	0	0	0	0	0	0.293	1.354
49	225	225	225	600	600	600	0	0	0	0	0	0	0.301	1.263
50	225	225	225	600	600	600	0	0	0	0	0	0	0.301	1.263
51	225	225	225	600	600	750	0	0	0	0	0	0	0.316	1.044
52	225	225	225	600	600	825	0	0	0	0	0	0	0.325	1.026
53	225	225	225	600	750	750	0	0	0	0	0	0	0.343	0.844
54	225	225	225	525	600	600	152	0	0	0	0	0	0.356	0.788
55	225	225	225	525	600	600	155	0	0	0	0	0	0.357	0.777
56	225	225	225	600	600	600	122	0	0	0	0	0	0.366	0.699
57	225	225	225	600	750	750	274	0	0	0	0	0	0.473	0.273
58	225	225	225	600	750	750	304	0	0	0	0	0	0.485	0.255
59	225	225	225	600	600	600	470	0	0	0	0	0	0.506	0.239
60	225	225	225	600	750	750	418	0	0	0	0	0	0.529	0.209
61	225	225	225	525	750	750	537	0	0	0	0	0	0.550	0.201
62	225	225	225	600	750	750	537	0	0	0	0	0	0.573	0.174
63	225	225	225	600	750	750	643	0	0	0	0	0	0.611	0.151
64	225	525	750	750	825	825	0	0	0	0	0	0	0.640	0.117
65	225	525	600	750	825	825	128	0	0	0	0	0	0.674	0.101

Sol#	Pipe Diameter (mm)						Storage (m³)						Cost (€x10⁶)	
	C	D	E	F	G	H	S1	S2	S3	S4	S5	S6	Investment	Damages
66	225	525	750	750	750	750	154	0	0	0	0	0	0.697	0.087
67	225	450	600	750	825	900	299	0	0	0	0	0	0.746	0.075
68	225	600	750	750	750	900	296	0	0	0	0	0	0.789	0.047
69	525	525	750	750	825	825	0	0	0	0	0	0	0.827	0.001
70	600	750	750	825	825	825	184	0	0	0	0	0	1.020	0

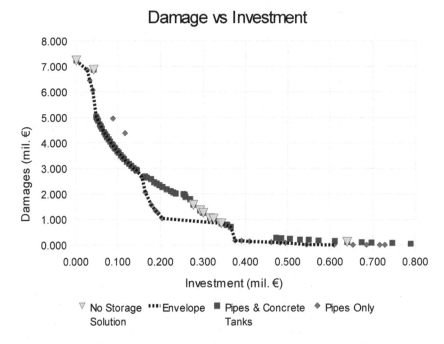

Figure 5.22: Comparison between the use of concrete tanks & pipes renewal against the use of only pipe renewal as rehabilitation technique

5.9 Intangible costs

The evaluation of intangible damages is subjective and very difficult to quantify (by definition). Anxiety was selected as the third objective to minimize. The quantification of anxiety is based on the methodology presented by Lekuthai and Vongvisessomjai (2001) that states that flood depths between 0.1 m and 1 m will gradually affect people. The scale is in terms of a percentage starting from 20% to 80%. Below 0.1 m people do not feel anxious. However, there is always a degree of anxiety if there are few centimeters of flooding (20%). The scale never reaches more than 80% because it is not tolerable for a human being. The intrinsic anxiety per person (A_0) is a function of the flood depth and the duration. Refer to Lekuthai and Vongvisessomjai (2001) for more detailed equations. It is assumed that anxiety is cumulative for the whole population.

Some additional terms have been incorporated into Lekuthai and Vongvisessomjai's method. A factor lambda (λ) was introduced in order to take into account of the mutual support between people. During flooding people tend to form groups and give some comfort to each other. This factor is set between zero and one (*0-1*). The factor depends on the population size at each node of the storm network.

Another factor to be considered is vulnerability (μ). Inside a group there are persons who are more vulnerable than others; for instance, old people and children. This factor must be greater or equal than one. A vulnerable person tends to get more anxious during an event.

An income factor (f_{inc}) was also used. People with lower income get more anxious than people with high income. If a person has more than one house the stress is less compared to a person with only one house. The income factor is assumed to be exponential and depends of two factors k and I (equation 58). The income is a value in the range between 0 and 2, with 1 as the pivot or reference value.

$$f_{inc} = e^{-k(1-I)} \tag{70}$$

where

f_{inc}	income factor
k	a factor greater than zero to scale f_{inc}
I	is the average income factor between 0 and 2

Finally, the inhabitants' past experience has to be taken into account. If an event of the same return period has happened in the past without major damage people will be less anxious. The opposite is also applicable: people who have lived through a bad experience will get more anxious. One factor f_{Tr}, dependent on the return period, can be defined and applied. The equation to estimate anxiety is as follows:

$$A = \sum_{node=1}^{nn} \left[A_0(\lambda^n - 1)/(\lambda - 1)\left[1 + n/m(\mu - 1)\right] e^{-k(1-I)} f_{Tr} \right]_{node} \tag{71}$$

where

A	is the accumulated anxiety
A_0	is the intrinsic anxiety for the node

λ	is a mutual support factor
μ	is a vulnerability factor
n	is the population in a specific node of the network
m	is the number of vulnerable persons in the node
nn	is the number of nodes
f_{Tr}	is a factor for previous experience (0 - 1)

5.9.1   Application to twelve pipes case

The case was applied to the same twelve pipe network with streets (see Section 5.7). The same objectives for cost investment, pipe renewal and damages cost were used in addition to the anxiety approach. After the application of the diagnostic tool to identify the critical pipes the same results were obtained as in the previous case, for exactly the same pipes. This is due to the diagnostic tool depending only on the relation Q/Q_f and not on anxiety. The values used for the computation of the anxiety for each node can be seen in Table 20. The optimization process using NSGAX tool was performed using a population size of 100 and run for a maximum of 50 generations.

Table 22: Parameter for anxiety computation

Node Id	λ	μ	m	I	k	Area (ha)	Inh./has
default	0.985	1.250	0.150	1.000	1.20	1.00	0
0014_1_c	0.985	1.25	0.20	0.90	1.2	1.00	100
0030_1_c	0.985	1.25	0.20	0.90	1.2	1.00	80
0078_1_c	0.985	1.25	0.20	0.90	1.2	2.00	70
0051_1_c	0.985	1.25	0.20	0.90	1.2	2.00	80
0073_1_c	0.985	1.25	0.20	0.90	1.2	1.00	100
K001_1_c	0.985	1.25	0.20	0.90	1.2	3.00	90
K002_1_c	0.985	1.25	0.20	1.20	1.2	5.00	90
K003_1_c	0.985	1.25	0.20	1.20	1.2	3.00	90
014_1_c	0.985	1.25	0.20	1.20	1.2	1.00	80
K006_1_c	0.985	1.25	0.20	1.20	1.2	5.00	90
K007_2_c	0.985	1.25	0.20	1.20	1.2	3.00	50

Node Id	λ	μ	m	I	k	Area (ha)	Inh./has
0127_1_c	0.985	1.25	0.20	1.20	1.2	2.00	20
Out_c	0.985	1.25	0.20	1.20	1.2	1.00	30

Analysis of Results

The maximization of the investment (pipe renewal) costs implies a reduction in the damage costs due to flooding, which is the expected behavior as shown in previous cases. There is a good agreement between flooding damage reduction and anxiety reduction, as is shown in Figure 5.8. This is because the anxiety depends on flooding depths. For this reason the Pareto shows a line in three dimensional space and not a surface, making the interpretation easier. However, when these results are compared with the results of the two objectives the initial problem (Section 5.7.5) it can be seen that there are more solutions with higher investment costs when anxiety is included in the objective function set. This means that more investment has to be made to improve the network system (see Table Error: Reference source not found and Table 21).

Table 23: Pipes diameter selection including intangible cost (anxiety)

Sol.	Pipe Diameters						Costs		
	C	D	E	F	G	H	Investment	Damages	Anxiety
1	225	225	225	225	225	450	0.000	7.240	15341
2	225	225	225	225	450	450	0.062	6.073	13340
3	225	225	225	225	450	525	0.106	5.667	12884
4	225	225	225	225	525	525	0.117	5.331	12507
5	225	225	225	225	525	600	0.124	5.119	12287
6	225	225	225	225	600	750	0.152	4.939	11998
7	225	225	225	225	600	825	0.161	4.966	11982
8	225	225	225	525	525	525	0.258	2.600	7656
9	225	225	225	450	525	750	0.259	2.273	7330
10	225	225	225	450	525	825	0.268	2.241	7244
11	225	225	225	450	600	750	0.271	1.807	6445
12	225	225	225	525	600	600	0.277	1.567	5814
13	225	225	225	525	600	750	0.293	1.345	5242
14	225	225	225	600	600	600	0.301	1.255	5053
15	225	225	225	525	750	750	0.320	1.078	4574

Sol.	Pipe Diameters						Costs		
	C	D	E	F	G	H	Investment	Damages	Anxiety
16	225	225	225	600	600	825	0.325	1.018	4369
17	225	225	225	600	750	825	0.352	0.839	3769
18	225	225	525	600	600	900	0.431	0.595	3362
19	225	225	600	750	750	825	0.513	0.252	2366
20	225	375	600	825	825	825	0.611	0.179	1719
21	825	1050	1050	1500	2000	2000	1.834	0.000	1190

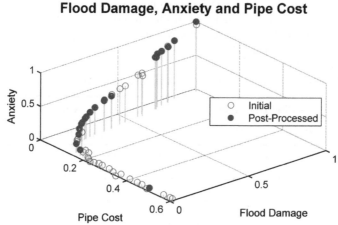

Figure 5.23: Normalized Pareto including intangible costs

5.10 Summary and Conclusions

This Chapter was aimed at describing the methods used to implement the proposed multi-tier framework for the rehabilitation of an urban drainage system, and perform a "*proof of concept*" for a small study case. Firstly, a structure for investment cost estimation was implemented; features like pipe replacement, storage tanks and ponds, and diversion structures were included. A method for damage cost estimation was also incorporated. The cost structures are based on equations in common use and from a unitary price analysis, including preset worth in cost estimation.

A prototype tool was developed and tested. A small study case was used as a "*proof of concept*". Two multi-criteria algorithms based on genetic algorithms were tested; they were NSGA-II and ε-MOEA. NSGA-II was incorporated onto a library and into an optimization tool called NSGAX. In order to compare the performance of the algorithms, four metric indicators of multi-criteria performance were used: cardinality, time, number of functions evaluations, hyper-volume and ε-Indicator. NSGA-II outperforms ε-MOEA in general, but the differences were not substantial. Hyper-volume and ε-Indicator show convergence, and they were used as stopping criteria during the optimization process. The application of the framework to a larger network with 63 pipes showed the need for more efficient algorithms or computers regarding computational time.

In both case studies of 12 and 63 pipes, it was possible to find rational solutions that may be expected by practitioners; for instance the storage tanks were properly selected depending on their location and were also satisfactory dimensioned. It is concluded that the developed framework is suitable for use in the rehabilitation of urban drainage system. It shows scalability and flexibility. The use of intangible costs was also evaluated through the implementation of an objective function to minimize the anxiety of the population. The tool allows for interpretation and negotiation between stakeholders over a set of alternatives and is not reliant in only one optimal solution.

6 Parallel Computing in Optimization of Large-Scale Systems

In this chapter parallelization of the computational processes in the developed framework presented in Chapter 3 is covered. Two approaches have been mentioned as suitable methods to deal with computationally demanding problems like optimization of urban drainage networks rehabilitation. One approach was through code improvement which was address on Chapter 4, and the other one relates to the use of parallelism in the code. Technological aspects of developing parallel code are reviewed. Parallel virtual machine (PVM) libraries were used for NSGAX parallelization, and the new MOSS algorithm was tested on an sub-catchments on Belo Horizonte city in Brazil.

6.1 Introduction

One of the main problems to face in the optimization of large-scale urban drainage systems is the computational cost. Solving the hydrodynamic equations in order to estimate water depth, velocity and flood duration can be a bottleneck in the optimization process. It becomes worse if dual modeling (using 1D and 2D-hydrodynamics model) is used. There are three main approaches to accelerate the computational process and improve the performance: first is to use a faster machine, second is to improve the algorithm, and the last one is to divide the whole process in smaller tasks and to use several computers or processors.

After some analysis of these options it was realized that the most suitable approach was the use of several personal computers (PCs) or processors. In the optimization process of a sewer network the most time consuming task is the evaluation of the objective function. As mentioned above the use of hydrodynamic models requires a high computation load. To improve the computation time through the use of software requires access to the source code and the most robust software codes are commercial and not open source. The use of a better and faster machine is a good option; however nowadays the PC is reaching its physical speed limits with current technology. Moore's law (1965) states that speed of the computer increases exponential in time, due to the use of smaller and faster transistors. This was true in the last 40 years but the miniaturization of the transistors is reaching the limit of the atoms and it is necessary to have a different technology to gain CPU speed. Nano-technology is promising in this aspect but is not sufficiently mature. Not only is the CPU speed a problem, but also the bandwidth and memory can become bottlenecks. For these reasons, CPU and computer manufacturers, INTEL and AMD, are moving to muti-processor computers (and multi-core processors) to gain computation performance.

The best action now to gain speed in computation is the use of multiple processors or the use of multiple computers. However, this is not straightforward; it requires for example techniques in parallel programming. The most suitable approach is the parallelization of the function evaluation in the multi-objective evolutionary algorithm (MOEA). The main goals of parallelizing MOEA are usually the acceleration of the computation and the gain of quality according to the approximation of the true Pareto front. Also, the diversity of the solutions would benefit through the use of parallel processing because it is possible to manage larger sets of alternatives than for a single optimization. In order to take real advantage of parallel computation, the serial algorithm has to be divided into smaller a number of serial tasks; such tasks must be independent or "quasi- independent" in terms of data and execution. This makes it possible to execute them in parallel using other processors or PCs.

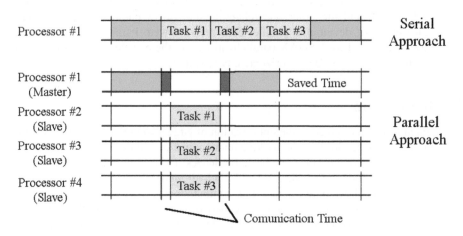

Figure 6.1: Parallel vs serial computing algorithms

Figure 6.3 shows an ideal scheme of a serial algorithm parallelization. A serial algorithm is divided into three tasks that are not dependent on each other, and each task is sent to an individual PC or processors, saving in this way valuable execution time. The total execution time is consumed in calculations and communications between PCs or processors.

6.2 Parallel Computing

There are different concepts that have to be addressed in parallel computing. Usually, there exists an ambiguity in the concepts which tends to create confusion among users and people who do not belong to the parallel computing community. There is ambiguity in the use of the term "*parallel*". There exist terms such as parallel computing, distributed computing, grid computing, cloud computing and other terms that refer to some kind of approach to the parallelization of a process or task. These terms depend on the physical (hardware) architecture, network topology, and the programming technique.

Hardware classification can be done in a number of ways. In the past it was common to talk about micro-computers, workstations, mini-computers, mainframes and supercomputers. This classification is not much used nowadays, and a more general classification has been substituted:

- Uni-processor computers: They use the basic structure described by Von Newmann, composed of a CPU, memory and a communication bus. The data and instruction are stored in the memory, transported by the bus to the CPU and instruction are processed by the CPU. The development of "*Reduced Instruction Set Computer*" (RISC) has reduced the cycle time per instruction allowing a speed gain in these single processors. Also, the inclusion of pipelines and super-scalar architectures has increased computer speed. However, there was not much improvement in the memory bandwidth and it is a serious limiting factor (Sloan 2004).

- Multi-processors computers: In this classification fall the computers from two to thousands of processors. The most common architecture is the use of centralized multiprocessors, also known as symmetric multi-processors (SMP). Depending on the memory management they are divided in uniform memory access (UMA) and non uniform memory access (NUMA). The UMA architecture shares the memory among the processors and in NUMA each processor has its own memory. Due to the memory bandwidth, the UMA architecture tends to be slower due to memory synchronization. Other multiprocessor architectures include vector computers. These process the same instruction in an array of processors; however they are not suitable for all types of problem configuration.

- Multi-computers: While in single and multiprocessor architectures the processor resides in the same machine sharing the rest of resources; for multi-computer architectures the processors are spread between different machines and each processor has its own resources. It is into this classification that the clusters fall.

6.2.1 Cluster, Grid and Cloud Computing

There is ambiguity in the use of the cluster, grid and cloud computing terms. Indeed, they are not more than instances of the same class, in which some properties are different. Cluster is the oldest term, it can be traced back to the mid-1970s with the HYDRA project (Wulf et al. 1974). The development of the multiprocessor operating systems and libraries like "*Parallel Virtual Machine*" (PVM) and "*Message Passing Interface*" (MPI) gave birth to the current "*Beowulf*" cluster. Grid computing is no more than a cluster that can be accessible from the Internet and their resources are not physically located at the same geographical place. Grid computing is aimed at sharing resources (storages, processors, supercomputers, PCs, PDAs, mobiles, etc.) through Internet (Myerson 2009). Grid computing was born in the 1990s within the scientific community; examples are the SETI@HOME project (Anderson et al. 2002), and projects from the European Organization for Nuclear Research (CERN) with the development of OPENLAB. Some commercial companies have called this type of computing "*Utility Computing*". Recently, the term "Cloud Computing" has been introduced. The main difference is that while the grid was designed to solve one particular problem, the cloud is aimed at running several tasks at the same time, and a task is not necessary to be one with the other (Velte et al. 2009). Cloud computing is composed mainly of thin clients and is intended to be a substitute the user's daily computer needs. The user does not own the hardware or software, he/she just leases it.

Cluster Topology

The most simple cluster topology is the symmetric cluster which is based on single nodes (Figure 6.2). It is very easy to set up on a sub-net with the individual machines and to install the required software. This system can present problems in security; for this reason in places where security is important asymmetric clusters are preferable. Additionally, a central (head) server is required for the computer nodes. It will solve the security problems working as a firewall. Also, the server can be used as an application and data server, allowing the nodes to be minimalistic regarding software. The drawback is when the cluster starts to grow and the server is not able to manage the load. In such cases the use of more than one server is recommended (*Expanded Cluster*). The application (IO/Server) and data (NFS) server can be separated; however this will increase the maintenance requirements in the cluster. The nodes in the cluster can be homogenous or heterogeneous in hardware and software.

Cluster planning

Building a cluster is a process that has to be planned. The following steps are suggested for building a cluster (Sloan 2004):

1.   Determine the overall mission of the cluster.

2.   Select the general architecture of the cluster.

3.   Select the operation system and software requirements.

4.   Select the hardware of the cluster.

The mission of the cluster is the most important because the rest of the steps in the design depend on it. Issues like who is going to use the cluster, how much load it will manage, what kind of software will be used, the availability of resources, access and security concerns have to be addressed in this step. The architecture will depend on the resources available. Whether, it is expected to lease a grid or it will be built from scratch. One of the main advantages is the possibility of using the existing installed network and equipment, which in the worst case will be a symmetric network. Different operating systems have different level of cluster support. Linux has been supporting cluster libraries for a long time, and has the advantage of being free open source. Windows has more recently supported software with parallelization purposes and has the advantage that most commercial and scientific software is developed to be used on this platform. Linux has well tested libraries for parallel programming: PVM and MPI are the most typical examples. For Windows the libraries are much more reduced, and PVM and MPI have been ported only recently. In the present research some libraries were tested but there were some bugs that make their implementation difficult.

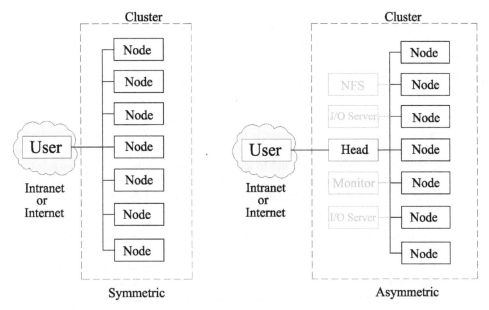

Figure 6.2: Cluster topologies

6.2.2 Concurrency

Two events are concurrent when they occur in the same time interval, but not necessarily in the same instant. For instance, two tasks may occur in a second but they may be executed in different fractions of a second. Single processors cannot execute instructions at the same instant, but through the operating system they can assign time to each task. Both tasks are being executed at the same time but not in the same instant. If more processors are available, the tasks can be executed at the same time. It can be said that both approaches present concurrency.

Concurrency is not managed by the processor itself: it is managed through software. Before the explosion of multi-core PCs less than ten year ago, the use of operating systems that could manage multi-processor computers was not wide spread. In the last ten years almost all of the operating systems are able to do it. However, not all software is able to take advantage of the multi-core features, and there is not much gain regarding the advantages of concurrency.

6.2.3 Parallel and distributing programing

There are several approaches to achieve concurrency.   One is the use of parallel programming and the other the use of distributed programming. Parallel programming techniques take advantage of multiple processors contained in one PC to execute one task while distributed programming uses several machines or a cluster to carry out the same task. Other authors state that parallel programming is the capacity to achieve concurrency in multiple processors in real or in a virtual machine. In this sense distributed programming is a sub-set of parallel programming (C. Hughes and T. Hughes 2003). This last concept is the one that is used in this thesis.

Parallel programming or distributed computing has to go through a design process. It has to take into account three main issues:

- Decomposition: the whole computation process has to be divided into smaller tasks in that parallel programing is based in the *"divide and conquer"* principle. The decomposition can be done in several ways; there is no straightforward procedure. The programmer has to identify the areas of the program suitable to be parallelized; not all problems can be parallelized and it should not be forced. One aspect to be careful of is the granularity. This represents the smallest part in which a piece of software can be divided.

- Communication: all sub-tasks that were decomposed need some sort of communication. They have to use memory and/or resources, identify if other sub-tasks have finished their jobs and collect results. In order to accomplish such tasks they need to communicate with a master or the operating system. The programmer has to be aware that such communication tasks will consume time and resources.

- Synchronization: usually, the created sub-tasks are not totally autonomous. They dependent on the master task or the head that will collect the results that form the final solution to the problem. The master has to be in charge of the coordination which is executed first between sub-tasks, which can be in parallel, and the access to resources like memory, disk storage, network, etc.

Layers of concurrency

There are two main layers in which concurrency can be applied. One is the hardware/firmware layer and the other is the software layer. As was mentioned above, the computer can contain single or multiple processors. The last generation of single processors with RISC architecture allows the execution of two or more instructions at the same time, speeding up the computation; this can be considered a kind of hardware concurrency. Also, in multiprocessor computers each processor is RISC based giving a level of hardware concurrency. A second layer of concurrency is managed by software. The operating system is the first software layer to implement concurrency; it allows concurrency through *multitasking* and *multi-threading*. In the case of single processor machines this concept allows for concurrency but not necessarily a speed gain. The speed gain is just because the processor is idle for shorter periods. In the case of multiprocessors, the operating system is able to select to which processor the instructions of each task or thread are sent, allowing the execution of tasks at the same instant. It balances the processor load using a threshold: when a processor is saturated with a large number of instructions the new tasks are sent to the other processor. However, this balance is not usually optimal.

Another layer of software concurrency can be implemented in applications. Because operating systems are not optimized to provide parallelism, our application has to be designed to use the advantage of multi-processor or multi-computer architectures. Concurrency can occur at four levels:

- Instruction Level: this can be achieved when multiple parts of a single instruction can be executed at the same instant. This type of parallelism is supported by compiler directives.

- Routine level: also the routines, functions or procedures can be executed with concurrency if they do not depend on each other. They are launched as new tasks in other processors. Their implementation is done through the use of specialized libraries or multi-thread features in the operating system.

- Object level: this is a particular case of the routing level, but deals with the encapsulated object. A common implementation is done using the Common Request Broken Architecture (COBRA).

- Application level: more than one application can be run in parallel, on different computers or assigned to run on other processor, to work cooperatively among them.

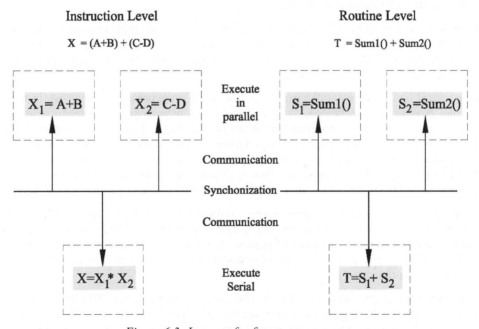

Instruction Level

$$X = (A+B) + (C-D)$$

Routine Level

$$T = Sum1() + Sum2()$$

$X_1 = A+B$ $X_2 = C-D$ Execute in parallel $S_1 = Sum1()$ $S_2 = Sum2()$

Communication

Synchonization

Communication

$X = X_1 * X_2$ Execute Serial $T = S_1 + S_2$

Figure 6.3: Layers of software concurrency

Implementation of concurrency at instruction level is achieved through compiler directives but not all the compilers support parallelism. There exist several compilers which have been developed to be used in super computers, on hundreds or thousands of processors, like *Parafrase*, *Bulldog* and *High Performance FORTRAN* F90 among others. All of them are aimed to compile FORTRAN source code into parallel executable code. In the last decade C and standard FORTRAN have been adopted for PC architectures, the so called OpenMP standard. It is a standard to compile parallel programs at instruction level. Currently, OpenMP is supported by INTEL C++/FORTRAN (v. 10.1), Gcc (v. 4.3.2), Sun Micro-systems and IBM compilers among others (Chapman et al. 2007).The main advantage is the use of shared memory - it allows the variables to be addressed directly with less time consumed.

Concurrency at instruction level is suitable for several processors in one physical computer (multiprocessor computers), but it is very difficult to implement on multi-computer architectures (clusters). In such cases the use of distributed programming is preferable; this implementation is done at higher level, routine or application level, through the use of Application Programming Interface (API) libraries. These libraries are based on the concept of message passing between applications in order to share data. This is because each sub-task has their own memory and data space. There are two main standard libraries for parallel programming for distributed computing, one is the *Parallel Virtual Machine* (PVM) and the other is *Messaging Passing Interface* (MPI).

Granularity and Communication time

Granularity is related to code size and the relation between the proportion of parallel and serial code. When the code size is small and the execution time is short, then the algorithm can be defined as "*fine-grained*". The opposite, namely a large time of execution per task, it is called "*coarse-grained*" parallelism.

One critical issue influencing the parallel performance is the communication cost. In most parallel programming techniques, it is necessary to stop computing in order to send and receive messages (data or command) between parallel tasks. Theoretically, the more the number of parallel tasks implies a better performance in the parallel algorithm. However, if communication time between processors is taken into account, this statement is not totally true because when the task execution time is less than the communication time, the total performance of the parallel algorithm becomes worse than a serial algorithm.

"*Fine-grained*" algorithms require more communication between processors while "*coarse-grained*" need less. However, the last one has more serial code and this makes it more inefficient even if the number of processors is increased. In other words, the parallel code must be as "*fine-grained*" as possible but without reducing performance due to the communication time.

A combination of programming APIs is also possible. In particular, MPI or PVM with OpenMP may be used together in a program see (Mahinthakumar and Saied 2002), which may be useful if a program is to be executed on a cluster that consists of multi-processor architectures. The main reasons for doing this is to take advantage of a finer granularity of parallelism than possible with any parallel API, reducing memory usage or reducing network communication (Gropp et al. 1994).

6.2.4 Parallel Virtual Machine

Parallel Virtual Machine (PVM) is one of the API libraries that can be used in distributed programming. It dates from 1989 and was developed by the University of Tennessee, Oak Ridge National Laboratory and Emory University. Originally developed to be used in UNIX-like operating systems it has been ported to other architecture as well. Originally it was developed to perform parallel computation on heterogeneous clusters, with single processors to multi-processors and even supercomputers, through a passing messages methodology. It makes it possible to reassemble the cluster of computer like a unique virtual multiprocessor computer, in which each individual computer is one processor. PVM API libraries show the following characteristics:

- Heterogeneity in term of machines, networks, and software

- Explicit message-passing model

- Process based computation (at routing and application level)

- Single and Multi-processor support

- Transparent access to hardware

- Scalability, allowing for dynamic configuration. It is possible to add or delete nodes (computers or virtual processors) at run time.

The library is composed by two main parts: the library itself and a server or daemon. The PVM daemon (pvmd) serves as a message router and controller. It provides a point of contact, authentication, process control, and fault detection. The pvmd occasionally checks that its peers are still running to aid in debugging. It is in charge to start the other daemons on node computers. The pvmd that starts the tasks is named the master and the ones that are started by it are the slaves. The library (pvm3.a or pvm3.lib) is the API itself. The developed parallel code has to be linked to this library in order to use the function to be used in the parallelization.

Some other tools are also available in the package. One is the PVM console and it is used to start the master pvmd daemon and start the slave nodes by the user. It also allows to start the task and keep track of the task sent to each node. From this console also the task can be aborted if needed. Another important tool is XPVM. This is a GUI for the PVM console but additionally it can monitor the activity in the master computer and the nodes. It allows to delete and add nodes dynamically, monitor the messages between nodes, visualize if one node is working, waiting or idle in function of time.

The PVM library was developed some time ago but it still is in wide spread use and mature (Geist 2005). It is maintained regularly adding new features. The library can be linked with any ANSI C, C++ and FORTRAN. It has been also ported to be used with OCTAVE, MATLAB, SciLAB, R and other environments. PVM is the easiest to use and the most flexible environment for basic parallel programing on heterogeneous clusters: some authors recommend it for small and medium size parallel programing tasks (Hughes and Hughes 2003) however, It also has be applied to complex and large problems - see (Fang et al. 1999), (Hluchý et al. 1999), (Ouenes et al. 1995), (Jiang et al. 2005) among others.

6.2.5 Messaging Passing Interface

The Message Passing Interface (MPI) is a standard. Before MPI each manufacturer of supercomputers or distributed memory computers had their own way of establishing communications between processors and data segments. This makes it very difficult to port applications developed on such machines to other machines with a similar architecture, or even more difficult onto computers without distributed memory architectures. In order to solve this issue, in 1992 vendors started to unify to a single standard the required set of instructions to use memory and processors in multiprocessor computers. MPI development started with a workshop sponsored by the Center for Research in Parallel Computing involving vendors like IBM, INTEL Nx, nCUBE Vertex, PVM developers and others.

The main goal was to create a standard to build programs for a distributed memory communication environment with the following features:

- Design of an application programming interface rather than for the compiler

- Allow for an efficient memory management, avoiding copying memory to memory and direct communication among processors.

- Allow heterogeneous environments

- Allow C and FORTRAN programming

- Interface is like PVM and allows flexibility

- It can be implemented on many vendor's platforms

- The semantics of the interface is language independent

- The interface is designed to allow for thread safety.

MPI has two version: MPI1 and MPI2, the latter is the current standard. There exist many implementation of MPI; remember that MPI is a standard and not a library. The most common implementation in use on clusters are MPICH, LAM/MPI, and Open MPI. In order to build your own parallel programs using MPI at least one implementation has to be installed. MPICH like PVM is composed of one library and one daemon. The way to build your own parallel programs is similar to PVM. PVM and MPI has similarities but also differences, for a deeper analysis refers to (Geist et al. 1996) and (Gropp, Lusk, and Skjellum 1999)

6.3 Performance Measurement

Due to the existence of different algorithms, methods, architectures and topologies to carry out parallel computations, a suitable, easy and reliable method to compare performance is required. The use of computers carries a cost. This cost can be due to depreciation in time of the computers, electricity consumption, maintenance etc. The addition of a new computer to a cluster implies an increase on these costs. If the gain in execution time (or money) is not proportional to the inclusion of such new equipment it means that it would be preferable not to increase the number of PCs in the cluster. There are several benchmarks to check the performance of the network; most of them are based on the execution time. However, "*Equal Duration*" and "*Parallel Computation with Serial Sections*" models are the two methods mainly used.

Equal Duration Model

This model assumes that the total process can be divided into n equal tasks, and each task can be sent to and executed by one processor (or PC). If t_s is the total time consumed by the process to accomplish the whole task in a single processor, it can be said that each parallel task theoretically takes $t_p=t_s/n$ to be executed simultaneously. However, this is only in theory, because when a parallel approach is used there are some time losses. The speedup is defined as the ratio between the time consumed to solve one problem (process) using a serial approach in a single processor (t_s) and time used to solve the same problem in n processors (t_p) working in parallel (Eq. 72). If $t_s=n\,t_p$ means that the speedup $S(n) = n$, such that the speedup depends only on the number of processors, the more processors the greater the gain.

$$S(n)=\frac{t_s}{t_p}=\frac{t_s}{t_s/n}=n \tag{72}$$

However, this statement misses an important part of the parallelization process, which is the communication overheads, namely the communication itself and the data exchange between tasks. If we call t_c the time employed in communications for one processor, such that $t_p = t_s/n + t_c$ then Eq. 72 can be re-written as shown in Eq. 73 and normalized as in Eq 74. This normalized form is called the speedup efficiency.

$$S(n)=\frac{t_s}{t_p}=\frac{t_s}{t_s/n+t_c}=\frac{n}{1+n\dfrac{t_c}{t_s}} \tag{73}$$

$$\xi=\frac{1}{1+n\dfrac{t_c}{t_s}} \tag{74}$$

where:

$S(n)$ is the Speedup with communication time

ξ is the Speedup efficiency

t_c is the communication time

t_s is the serial time

It can be seen that when $t_c \ll t_s$ the Speedup tends to n. When $t_c \gg t_s$ the Speedup tends to be less than one, and communications becomes more important.

Parallel Computation with Serial Sections

This model considers, additionally to communication time issues, the fact that it is impossible to split the whole process into parallel tasks. For instance, a task used to compute the addition of two vectors can be parallelized sending one sum to each processor, but a task to sum all the vectors values can only be executed in serial order (Figure 182). This means that some serial code fraction (f) will always remain, and the time to execute n tasks in parallel can be estimated as $t_p = f \cdot t_s + (1 - f) \cdot t_s / n$ leading to Eq. 75 for the speedup.

Parallelizable Code	No Parallelizable Code
For i=1 to n	*For i=1 to n*
$\quad c(i) = a(i) + b(i)$	$\quad a(i) = a(i) + a(i-1)$

Figure 6.4: Example of parallelizable and no parallelizable codes

$$S(n) = \frac{t_S}{f \cdot t_s + (1 - f) \cdot t_s / n} = \frac{n}{1 + (n-1) f} \qquad 0 \leqslant f \leqslant 1 \qquad (75)$$

If the value of f is zero, that is, there is no serial fraction, the Speedup tends to be equal to n while for a value f of one the Speedup is equal to one, meaning that we have a full serial code. On the other hand if n is equals to one, that is, there is only one processor, the Speedup implies pure serial code. When $n \to \infty$:

$$\lim_{n \to \infty} S(n) = \lim_{n \to \infty} \frac{n}{1 + (n-1) f} = \frac{1}{f} \qquad (76)$$

This equation shows that no matter the number of processors, there will always be a limit to the gain in the parallelization of the whole task, which depend on the intrinsic serial fraction of the algorithm. This last statement is known as Amdahl's law (Amdahl 1967). If communication time is considered the following equation is used:

$$S(n) = \frac{n}{1 + (n-1)f + n \cdot \dfrac{t_c}{t_s}} \tag{77}$$

Amdahl's law is somehow pessimistic in stating that it does not matter what number of processors are used. In practice it has been observed that some problems have shown a behavior of Speedup that is almost linear. It usually happens in scientific and engineering problems in which the parallel part of the algorithm scales up with the problem size. This issue was addressed by Gustafson (1988) who proposed Eq. 76 as the performance equation. This equation takes into account the scalability on the size of the problem.

$$SS(n) = f + n \cdot (1 - f) \tag{78}$$

where: $SS(n)$ is the scaled Speedup factor with communication time

Another anomaly is the so-called super linear speed-up, which means that the speed-up has been measured to be more than n. This may happen because of memory access and cache mismanagement or because the serial implementation on a single processor is suboptimal (Karniadakis and Kirby 2003).

A drawback in the use of the presented equations is that estimation *a priori* of the values of the serial and parallel fractions is not a straight forward procedure. The speedup and/or the number of optimal processors have to be defined through an iterative procedure, and comparative scenarios have to be tested for the specific problems in order to arrive at a rule of thumb. In the case of urban drainage optimization combined with hydrodynamic models there are no references to the number of processors to use in order to be efficient in the use of resources.

6.4 NSGA-II Parallelization

The parallelization of the Multi-Objective Evolutionary Algorithm (MOEA) is intended to speed up the optimization process regarding computational time. MOEA can be parallelized in its genetic operands or in the objective function evaluations. The first has been addressed in multiprocessor computers with shared memory, in which access to the computer resources (memory, buses and CPU) is faster than in clusters. The second is used in clusters of computers where more time is consumed in communications making the parallelization of the genetic operands unreasonable but making possible the parallelization of the objective functions. In the case of urban drainage optimization, where hydrodynamic models are used to compute the objective functions, the second way is preferred. There are three main approaches or paradigms to parallelize the MOEA algorithms: Master-slave, Island, and Diffusion model approaches (Figure 6.5). There are more classifications, but almost all of them can fit into these three main models. The selection of one model or another depends on the nature of the problem.

In Master-slave models, one processor (master) is in charge of sending the data to the slave processors and collecting the results after execution. The only communications exist between the master and slave processors. MOEA genetic operators are executed in the master processor and the function evaluations in the slave processors. In the case of the Island model, the MOEA population is divided in sub-populations and a MOEA is run for each sub-population in its own processor. They will converge to different regions of the Pareto front. They share the good genes in order to build the whole Pareto set. The Diffusion model deals with one population based in a neighbourhood structure. The genetic operators are applied only between neighbours on the structure, and each individual is assigned to a processor. More detailed information about MOEA parallelization can be found in Coello et al (2002).

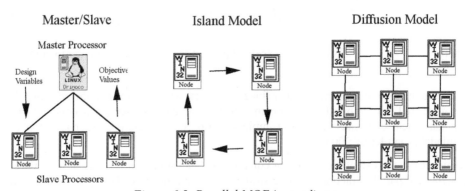

Figure 6.5: Parallel MOEA paradigms

6.4.1   NSGAXp algorithm

The multi-objective evolutionary algorithm selected for parallelization is NSGA-II. It was selected due to its well-tested robustness to converge to the Pareto front, and it has shown good performance as identified in the preceding sections and in Barreto et al (2006). The algorithm was parallelized in a "coarse-grained" fashion, which is more suitable for a Master – Slave model.

The newly developed algorithm is called NSGAXp. It was developed in C++ and compiled using GNU g++ and Delphi, taking as its base the serial version for windows NSGAX by Barreto and Solomatine (2006). Due to the bottleneck being the hydrodynamic model, the task suitable for parallelization is the individual evaluation, in other words the objective evaluation which involves running the hydrodynamic model.

The program is divided into two algorithms, one that works as master, called NSGAXpm and the other that performs the objective function calculations, called NSGAXps. In between there is interface layer composed of a scheduler and parallel library to be used. Figure 6.6 shows the flowchart schema of the parallelization approach. The importance and design of the scheduler is below.

The master, NSGAXpm, is in charge of controlling the evolutionary operand, and the following set of instructions are needed to achieve parallelization:

1. Start the slave nodes: this command uses the daemon (server) to start all the nodes that form the parallel virtual machine. This process can be done also manually, but the user needs physical access to each node, making it impractical if there are more than 2 nodes.

2. Send a configuration command: The master indicates to a slave to carry a set of predefined instructions that maybe necessary before the function evaluations start. For instance, if more that one function evaluation task is sent to the same node or processor, the slave has to open independent folders for each task in order to avoid mixing result and errors due to accessing the same files.

3. Send individuals for evaluation: The master has to send each individual of the population to each slave task opened by the master in order to be evaluated and wait for the results of each slave.

4. Apply the genetic operators and repeat step 3 until the stopping criteria is met.

5. Send a termination command to each slave in order to free memory, delete temporal folders and files, and close the slave.

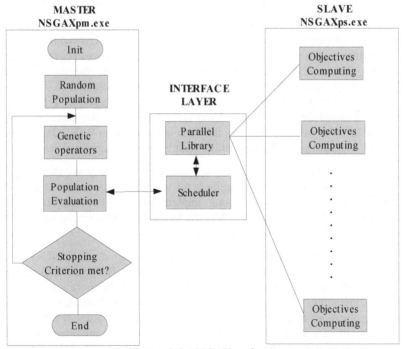

Figure 6.6: NSGAXp schema

The slave, NSGAXps accepts the following set of instructions from the master:

1. Waiting loop: Once started the slave waits in a loop for a command from master:

    *   Configuration: As it was stated, this command executes a set of instructions needed before evaluating any individual from the population

    *   Evaluate individual: If the command is an individual evaluation, the slave starts a receiving process. It receives the design variables to compute the objective functions. This includes running the hydrodynamic model.

    *   Termination command: It ends the task on the slave computer

2. If the command is different to a termination command, the algorithm returns to step 1, for the waiting process.

6.4.2 Library and OS Selection

Two main platforms exist for implementing parallel algorithms: supercomputers and cluster of PCs. Super-computers are faster due to their built-in components such as processors, memory and communication buses but they are expensive, and usually only available for major universities and research institutions. On the other hand, PC computation power is increasing and getting cheaper each year, and it is common to find several PCs in consultants' offices, universities, public service offices, etc. The total computation power is under-utilised; for example, during the night time PCs are switched-off wasting around 66% of the installed computational power.

The approach introduced here is aimed at being used in low cost environments as exist in developing countries. It is very common to find several alternatives regarding Operating Systems (OSs), Microsoft-Windows, LINUX and MacOSx are the most frequent. In developing countries, open source is desirable but not always possible, so it is imperative that the approach adopted here can work in multi-platform environments. For instance it is common to have mixed OS, LINUX and Windows; NSGAXp master algorithm can run in LINUX but MOUSE or MIKE 21 do not; they can only be run in WINDOWS OS. So the NSGAXp slave was designed to be compatible with WINDOWS applications.

Current PCs follow a sequential processing approach. In order to make them work like a parallel machine, a software layer is required. As was stated above, there are two main libraries that allow for parallel processing implementations: Parallel Virtual Machine (PVM) and Messages Passing Interface (MPI). PVM and MPI have similar applications and there is still ongoing discussions on which of them has to be used.

MPI has a very good performance on large multiprocessor machines, it contains a very rich set of functions for communications, and a high performance. But when applications must be executed on heterogeneous workstations, PVM appears to be more convenient. It has very good interoperability between different types of hosts and allows the development of fault tolerant applications that can continue despite errors in any guest or task (Geist et al. 1994). The concept of a virtual machine that implements PVM provides a basis for the heterogeneity and portability. More detailed comparison between PVM and MPI can be found in Geist et. al (1996) and Gropp and Lusk (2002). The parallel algorithm presented in this paper was developed to run under distributed environments both homogeneous and heterogeneous, it was then implemented in PVM 3.4. Also in PVM the implementation using master-slave model is easy. A good management of errors capture is available for PVM.

6.4.3 Network Topology Selection

The cluster to test our parallel implementation was configured as in a normal office network, and it was added to the existing UNESCO-IHE network. This was done intentionally because it is not always possible to optimize the network connexions at offices and the aim is to carry out the test over a normally configured network. The network configuration issues such as network topology and components for fast computation (Gigabit networks), which are very important for "fine-grain" cases, are beyond the scope of this study.

A set of ten old (and cheap) computers was selected as identified in Table 1. Note the heterogeneity in brand, speed and operating system of the network. This makes it more difficult to estimate the performance, but it represents a common office network configuration. All Ethernet network cards were 10/100 MBps connected with two switch HP 212M at 10/100 MBps. Figure 5 shows the network configuration, topology, OS, function and PC number. Additionally, the main institutional network is composed of more than 100 PCs which increases the communications on the network and introduces more noise to the experiments. This main network has a file server and a licence server. One issue is that the hydrodynamic model has to check the license over the network, which increases communication time and the computational delay of the objective functions evaluation.

Table 24: Network PC nodes configuration

Qt	CPU	RAM KB	OS	Speed MHz
1	AMD Atlhon (Laptop)	128	Debian LINUX etch	800
4	Intel Celeron (Desktop)	128	Win-2000	1800
4	Intel Celeron (Desktop)	128	Win-2000	2000
1	Intel P4 (Desktop)	256	Win-2000	1800
1	Intel P4 (Desktop)	256	Win-2000	2400

6.4.4 The Scheduler

One of the most critical parts of parallel computing algorithms is the way in which the load on the processors is managed. This kind of managing resources is termed scheduling, which can be understood as allocating resources over time to perform tasks that are parts of processes (Blazewicz et al. 2000). The scheduler must distribute the jobs to the processors, and control the access to shared memory and disks in an optimal way. The most simple scheduler is that used by the operating system, in which a threshold is set depending in the percentage of CPU use. Other common approach, which is the default in PVM and MPI, it consists on a round robin selection. A variant of such a method is the use of a weighted round robin selection. Other more elaborated approaches based on heuristic methods like GAs, ANNs or hybrid methods can be found in the literature see Elleuch et al. (1994), Moore (2004), Jin et al. (2008), Sudarsan and Ribbens (2010).

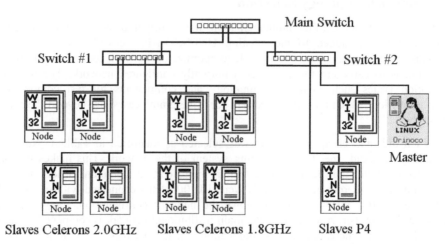

Figure 6.7: Cluster network topology

In the round robin approach the processors are virtually located in a pool and have equal probability to be selected on a random selection basis. Once a processor is selected it is taken out of the pool and the next request for a task is assigned to a new processor in the pool. Once a task is completed and a processor is free it is returned to the virtual pool list again. The drawback of such methods is that if the cluster is heterogeneous, and computers have different memory, speed, processors, it could be the case that tasks are sent to a slower processor leading to a poor performance. For that reason the weighted round robin is preferred. Such approaches are based on ranks: the virtual pool is ordered by a rank depending on the cluster components. Some of the parameters to be considered to build a weighted round robin scheduler are the following:

- CPU clock speed

- Number of processors per node (multiprocessor machines)

- Available memory (RAM) and disk

- Cache memory in processors

- Sub-network vs cards speed

- Load in the processors

- Number of tasks in the pool

- Physical distance between nodes

In our case, the scheduler was build using a round robin procedure, taking into account CPU clock speed, tasks in the pool to be processed and the number of processors. The nodes (PCs) of the cluster are dynamically ranked by relative speed, those nodes with multiprocessors are weighted equally but multiplying by the number of processors (Eq. 77). This is done in order to reduce the communication load in the network. If the number of tasks waiting to be executed is greater than the number of processor in the pool the scheduler will wait until a processor is free to send the next task.

$$f_{Speed} = \frac{CPU_{speed}}{Reference_{speed}} \cdot np \qquad (79)$$

where:

f_{Speed}	is the relative speed
CPU_{Speed}	is the CPUs Speed in MHz
$Reference_{Speed}$	is the Speed of reference in MHz
np	is the number of processor of the node

6.5 Parallel Computing Application

There exist a few applications of GA algorithms optimizers in sewer networks, but none have been found using clusters for parallel computing. Through the use of parallel multi-objective algorithms, it is expected to improve the performance and to reduce the time required to optimize a sewer network. However, one of the most important aspects is to know the optimal size and performance of the cluster. As was mentioned above it will depends on the granularity, communication time and intrinsic serial code. In order to evaluate the performance and the effects of the cluster size, two study cases were used in applying the parallelized algorithm. One is the previous case with twelve pipes, and the other a section of a network located in Belo Horizonte, Brazil.

6.5.1 An Small Study Case

This study case is a simplified pipe network composed of twelve pipes, thirteen nodes and eleven sub-catchments used previously. The objective functions to evaluate are the investment cost due to pipe renewal and flooding damages. The genetic algorithm for the study case was set-up using a population size of 100 individuals and running for a maximum of 50 generation. The crossover probability was set to one (1), and the mutation probability was chosen to be 1/np, where np is the number of pipes to optimize. Pipes were selected from a catalog of commercial pipes that contains 14 pipe costs. In the optimization the objective functions were evaluated 5000 times.

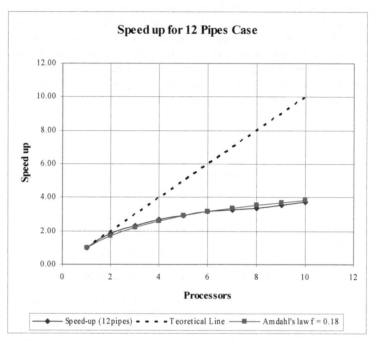

Figure 6.8: Speed up for 12 pipes study case and Amdahl's law

Using one processor, the total time was 12.78 hours. Figure 6.8 shows the Speedup from one to ten (10) processors, using the topology shown in Figure 6.7 and Table 24. The increase in performance when a processor is added can be seen from the plot. Adding more processors does not imply a linear speed increase, because part of the running time is used for communications and more processors implies more communications and more traffic in the network. Also if Amdahl's law is applied not all the code is parallelizable giving a non-linear behavior. The fraction of not parallelized code was determined using the Eq. 75 and 76, giving a result of 18% for Amdahl's law and 56% for Gustafson's law. It can be seen from the figures that Amdahl's law represents better the behaviour of the Speedup than Gustafson equation.

Using the completed cluster of 10 computers, the total time used by the cluster for the 5000 function evaluation was 4.77 hours, this implies a reduction of 8 hours in total was achieved representing a percentage reduction of 62.5 %. However, the use of ten processors perform like a five processors cluster as can be seen in the figures. It is true that there is a gain in computational time using more resources (CPUs or PCs), but this gain was not proportional to the processor investment. We can say that the use of ten processors will gave an efficiency of 50% in resource consumption. Regarding the optimization results, the pipe diameters and flooding costs obtained are in agreement with the results obtained in the previous chapter.

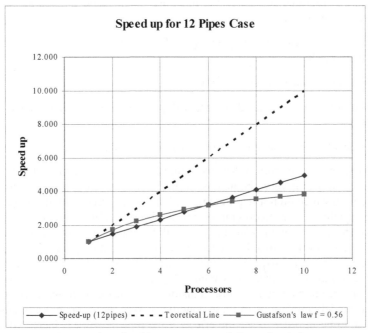

Figure 6.9: Speed up for 12 pipes case and Gustafson's law equation

6.5.2 Belo Horizonte Drainage Network

Belo Horizonte (BH) is the capital of the State of Minas Gerais, which in economic terms (gross product) is the third among the 26 Brazilian states. The city lies at 20° South latitude and 44° West longitude (Figure 6.9) and has an altitude of 750 to 1,300 meters. It is located in a mountainous region. Tropical highlands weather predominates in this area, with average yearly rainfall of 1,500 mm and average yearly temperature of 26°C. The rainy season lasts from October to March, when 90% of the total yearly rainfall occurs. The highest monthly average rainfall (315mm) takes place in December. Typical rainfall intensities are also relatively high (e.g.: 200 mm/h in the case of a 10-year return period event with 5 minutes duration; 70 mm/h for the 1h and 50-year return period event). Mean relative humidity reaches 50% during winter and 75% in summer.

Stormwater management has been entirely under the responsibility of the BH municipality since the city foundation. Traditional storm water systems prevail in the city, although experiences with detention ponds exist since the 50s. There are at about 4,300 km of roads all of them equipped with gutters, inlets etc. The municipal database on drainage infrastructure keeps details about 64,000 inlets (gullies), 11,500 manholes, 1,100 outflow structures (outfalls), and almost 770 km of stormwater sewers. There are some 700 km of perennial creeks in the municipal area. Part of those creeks have been lined, most of them as culvert concrete channels. The length of lined channels reaches near to 200 km.

The creek lining policy, which prevailed up to the 90s, was mainly justified by the following rationale:

- Lining is required for increasing the flow velocity and the channel conveyance, reducing the flood risk;

- Lining makes easy the implantation of interceptor pipelines and the so called sanitary roads;

- Lining makes easy the creek maintenance;

- Health risk due to directed contact with polluted waters may be reduced by creek lining;

- Inhabitants of riparian zones usually ask for creeks to be lined

The apparent simplicity of stormwater management, as perceived almost during all the last century, led to the use of very simple design methods for storm water systems. Synthetic models were used which do not require observed data to calibrate parameters (e.g.: rational method and synthetic unit hydrograph). Since observed data were considered as not necessary for storm water management, during all the last century the BH municipality did not invest in monitoring stream discharges or water quality parameters. One of the consequences of those approaches is high uncertainty in hydrologic design. A similar oversimplification also prevailed in hydraulic design. Complex flow conditions, including the effects of stream confluence, flow transitions or unsteady flow, were infrequently regarded and model simulations of these conditions were rarely done. Only uniform flow conditions use to be In order to see the effects if the pipe network is increased, a new test was carried out using

6.5.3   Building and Running The Initial Model

A portion of Belo Horizonte drainage network was setup in Mouse to be tested. The topography and the information to build the model were provided by the local partners in the learning alliance. The information was Pre-processed in ArcGis 9.0 to define the sub-catchments and main streams. This part of the network corresponds to the Vendanova Catchment located in the North side of Belo Horizonte. See Figure 6.13.

In order to see the effects if the pipe network is increased, a new test was carried out using a portion of the Belo Horizonte drainage network. The network is composed by 168 pipes and 169 nodes. A precipitation of 20 mm over a period of 6 hours was used. Flooding damage and pipe renewal cost are once again the functions to be evaluated. This time the pipe network does not consist of circular pipes alone, so the interface program was updated to manage circular, square and rectangular conduits. A new catalogue conduit catalogue was added for each type of section in the network.

Figure 6.10: BH location

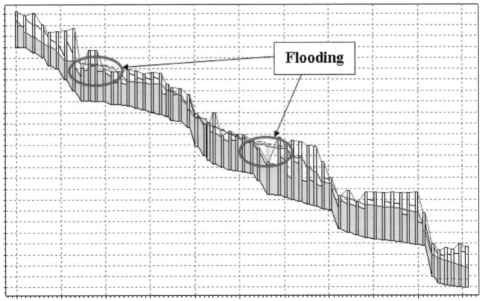

*Figure 6.11: Vendanova Catchment and
network layout*

Figure 6.12: BH flooded pipes

The initial run shows that there are some critical parts in the drainage network where most of the flooding occurs. Figure 6.1 shows the results of the initial run, the maximum flooding occurrence and a longitudinal profile of the main collector. For this rainfall event all the nodes that are flooded occur in the central collector of the system. The optimization process is setup to solve the flooding situation in the system with two objective functions, Flooding damage and pipe renewal cost. The total amount of pipes/channels that can be changed during the optimization process is 69. Each conduit (circular, rectangular or square) has on average 20 possible values in the catalogue. For this case the interface program was updated to manage circular, square and rectangular conduits. A new catalog conduit catalog is added for each type of section in the network. The total complexity of the optimization can be estimated as 69^{20}, which is 5.9836×10^{36} possible combinations. This large number of possible combinations is not possible to handle by applying the network changes manually. This is also one of the main arguments to use optimization techniques.

6.5.4 Belo Horizonte Analysis of Results

The network was run using the same approach as that for the previous case, but this time the population size was set 48 individuals and the number of generation to 200. The total number of function evaluations is 9600. Crossover is set to 0.9 and the mutation probability to 0.15. The optimization was done using up to six processors, but this time this cluster included two INTEL Pentium IV, and two INTEL dual core processors. The idea this time was to test other types of heterogeneity in the network topology, which includes the use of multiprocessor computers in the cluster conformation. The CPU speed for the Pentiums, and the core duo processors included were similar, 2.4 GHz each. The exercise was carried out 5 times.

A single run in a single processor took on average 81.54 hours, while for the whole six processors it took 16.10 hours. A reduction 80.3% in time was achieved, saving 65.5 hours for all the optimization process. Table 25 shows the performance comparison using Amdahl's and Gustafson's laws. This time the estimated fraction of non parallelizable code was 0.04 and 0.16 respectively. If we check for an efficiency of 50% as in the previous case, it can be seen that it is achieved using twenty processors. Once again, Amdahl's law performs better than Gusftason's, but this time both are very close to the real speedup. It can be seen that increasing the size of the problem, the sewer network in this case, improves the performance of the cluster – so advantages of. parallelism increase in case of solving large problems.

Table 25: Real and theoretical speed up comparison

Number of Processor	Speed UP	Amdahl's law (f=0.04)	Gusftason's law (f=0.16)
1	1.00	1.00	1.00
2	1.94	1.92	1.84
4	3.62	3.57	3.52
6	5.06	5.00	5.20
8	---	6.25	6.88
10	---	7.35	8.56
12	---	8.33	10.24

Number of Processor	Speed UP	Amdahl's law (f=0.04)	Gusftason's law (f=0.16)
14	---	9.21	11.92
16	---	10.00	13.60
18	---	10.71	15.28
20	---	11.36	16.96

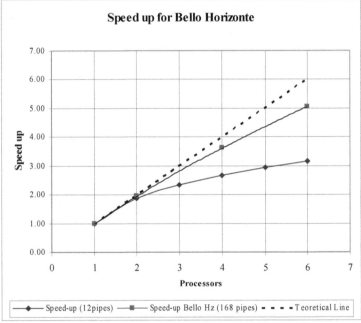

Figure 6.13: Speed up Belo Horizonte network

6.6   Summary and Conclusions

In this chapter it has been shown the state of the art regarding parallelism for computer code. Advantages and disadvantages of the different existing methods for parallel application have been studied. Theory of concurrency for multiple processors computers and cluster was reviewed. Implementation of parallel code for NSGAX was done using PVM libraries over Cygwin, a Linux OS emulator for Windows. The parallel code was included on the multi-tier optimization framework and applied to two case studies.

A small cluster composed by heterogeneous PCs with single and multi-core processors was set up. Two case studies were tested on the parallel framework: one using a small network with 12 pipes and the other for a sub-catchment in Belo Horizonte Brazil composed by 168 pipes. The results show good performance – saving between 60% to 80% of the computing time in comparison with single-computer optimization. Also, it can be concluded that the number of processor to use in the cluster has to be related with the size of the problem: sometimes it is better to use fewer processors. For example, when the 12 pipes problem was solved using 6 processor, the computing time was similar to the case of using only 3 processors, and for the Belo Horizonte case the use of 6 processors did not give any gain if compared to the use of 5 processors, but for larger problems advantages of using larger number of processors increase.

Chapter Seven

7 Framework Application in Developing Cities

This chapter has two applications of the multi-objective framework methodology developed in the previous chapters. The first application is to a city located in the west of Venezuela: Cabudare City. This city presents continuous problems due to the low capacity of the drainage system network which is exceeded by runoff events of rainfall of 1:2 years. Flood damages due to low return period rainfalls are low, but on the other hand the accumulated damages can be very high over a long time. For this real example the method has been slightly changed to use the expected value of damage cost, using several return periods of rainfall in stead of one single return period.

Another of the features that can be included and evaluated using the multi criteria framework is related to environment and water quality issues. The second application deals with three objective functions, facing investment cost, flood damage as a direct cost against water quality; thus including on the analysis a waste water treatment plant and temporary water storage. This was tested in the city of Cali located in southwestern Colombia. The implementation of both applications has not been straightforward due to lack of basic data. However the potential of the use of the developed multi-objective framework is illustrated for developing countries.

7.1 Expected Value Optimization – Cabudare, Venezuela

Cabudare is located in Lara State, Venezuela. It is the capital of the Palavecino municipality. The city has 119,671 inhabitants (from the census of 2001), which represents 44.4% of the total population of the Palavecino municipality. It is located between 69° 13' and 69° 18' longitude and 10° 03' and 10° 00' latitude, with an average elevation 400 m. The city is located in the "Rio Turbio" valley (Figure 7.1). The region is semi-arid; the average temperature is about 24 degrees and the annual precipitation is 464 mm/year. The city was founded around 1779. The old city center is located on the Tabure River. The valley around the old city used to be an agriculture zone with sugar cane as the main crop. Soils are heavy, composed of clays and silts. The area is very flat with slopes less than 0.1%.

7.1.1 Problem description

Due to the growth of the economy in Barquisimeto, the capital of the Lara state, Cabudare has become a dormitory community located less than 6 km of Barquisimeto, and new housing construction is replacing the sugar cane fields. A minority of the population is still engaged in agriculture within the municipality. The city has increased its urbanized area by more than 50 % in the last 20 years. In spite of low annual precipitation, rainfall intensities are high, as is usual in arid zones. These high intensities and the growing urbanization have increased flood damages. City urbanization was carried out using the existing storm network system without any upgrade. Nowadays severe flooding damage has been reported to buildings and infrastructure in the city, and more recently floods have endangered lives. Figure 7.2 shows one of the most significant recent storm events on October 16th, 2006.

There exists an urban master plan elaborated by the urban infrastructure ministry (MINDUR) for Barquisimeto and Cabudare dating from 1983. Also the environment ministry (MARN), based the plan prepared a drainage master plan for both cities in 1988. Regrettably, since that year to the present, Cabudare has developed in a different way than was planned; this has caused problems regarding the drainage.

The lack of a formal maintenance plan makes the situation more complicated. Maintenance is carried out on an ad-hoc basis and consists of replacing collapsed pipes, and cleaning channels and streams after damage has occurred. Few resources are assigned for such tasks and they are not used in an optimal way.

Figure 7.1: Cabudare city location and catchment delineation

The existing master plan (1988) was developed using the methodology established by the Venezuelan legislation. It consists of the selection of a fixed return period of rainfall to compute the peak discharge using the rational formula. Once the peak flow is estimated the hydraulic calculations are carried out using the Manning equation for steady state flow assuming that the pipe or channel will work under full capacity without pressurized flow.

Figure 7.2: Damages caused for the 2006 event in Cabudare

In order to have a better master plan, other more advanced methodologies, have to be used. Usually, the methodology recommended by the Venezuelan standard tends to oversize the drainage systems, wasting the scarce economic resource for investment in drainage. Another point to address is that the standard allows some inundation of streets in urban areas. However, the use of a steady state uniform flow does not permit the correct modeling and estimation of depths in the streets.

In the current methodology used in Venezuela, only the construction cost are taken into account; damage costs associated with the different events that occur the network are ignored. Finally, the possibility of storing storm water in order to reduce the damages in the urban areas can also be evaluated. In order to take these points into account new approaches are required which, in particular model the resulting flows from rain events by solving the full hydrodynamic Saint-Venant equations. Also, economic resources have to be optimized in a multi-criteria framework.

7.1.2 Data Collection

In order to have a reliable optimization process a hydrological and hydrodynamic model has to be built to compute water depths and velocities. In order to select the most suitable methodology the first step is to collect the available information. It was done through a visit to the different institutions in charge of measuring and storing rainfall data, through maps, standards, previous studies and personal visits to the study area. The available data and information that was collected is as following:

- Aerial survey scale 1:25,000 (source: Cartografía Nacional - 1988)

- Aerial survey scale 1:5,000 (source: MINDUR – 1989)

- Topographic survey scale 1:1000 (source: Antonio Rodriguez – 2006)

- Master plan for Barquisimeto-Cabudare (Source: MARNR 1988)

- Hydrological study of Cabudare metropolitan area (source: MARNR-MGA 1989)

- Precipitation at Cabudare station (source: MARNR 2008)

- Drainage Network System for Cabudare city (source: various)

7.1.3 Cabudare Hydrology

The next step was to build the intensity-duration-frequency curves from the collected raw rainfall data. The data consist of 25 years of extreme values for durations of 5, 15, 30, 60, 120 and 180 minutes. This data was used to provide a probabilistic distribution of rainfall for Extreme Values Type III (see Chapter 2) in order to produce Table 26 and Figure 7.3.

Once the IDF curves were computed the next step was to delineate catchment area as shown in Figure 7.1. The total area which contributes to the runoff is of 20.2 km². From a extreme event held on May of 1978 an areal reduction factor was calculated with a value of 0.888. Also, the concentration time was estimated using Kirpich equation (Chapter 2). The area has a larger part which is semi-rural and this equation has proven to work well in such a type of area. The total concentration time is 100 minutes. This concentration time was used to calculate the design rainfall. Six synthetic design storms were used. They were determined using the Chicago Method and by multiplying the resulting hyetograph by the areal reduction factor. The synthetic rainfall hyetographs are shown in Figure 7.4.

Table 26: Intensity-Duration-Frequency for various return periods and durations

Duration	Return period (years)					
(minutes)	2.33	5	10	25	50	100
5	270.0	323.3	363.3	416.7	453.3	493.3
10	235.0	281.7	318.3	365.0	400.0	433.3
15	208.9	250.0	283.3	326.7	357.8	388.9
30	156.1	188.9	215.6	248.9	273.3	297.8
60	105.0	128.1	146.7	170.3	187.5	204.7
120	64.0	78.8	90.6	105.6	116.7	127.8
180	46.3	57.1	65.9	77.0	85.3	93.4

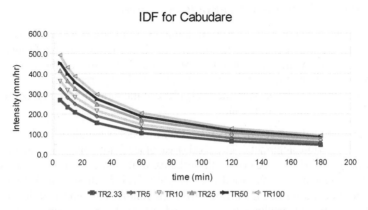

Figure 7.3: IDF curves for Cabudare rainfall station

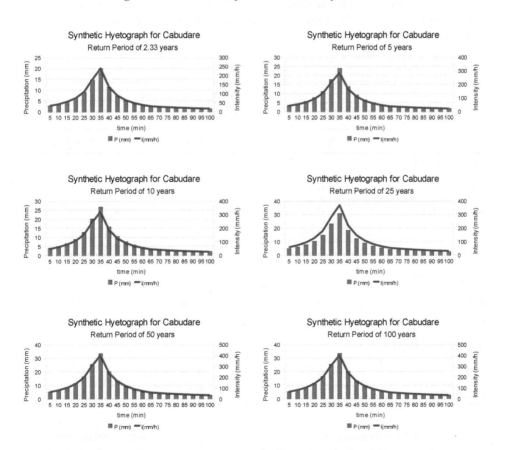

Figure 7.4: Synthetic hyetograph for TR2.33, TR5, TR10, TR25, TR50 and TR100 for Cabudare rainfall station

7.1.4   Model of Cabudare Drainage Network

Cabudare drainage network is composed of a main natural stream called Tabure River and an underground separate pipe network system which discharges to the Tabure River. As mentioned above the Venezuelan standard recommends the use of the rational formula. However for this study the method to generate the hydrograph used is the time-area method. This method has its basis in the coefficient of the rational formula, but it allows for a time distribution of the precipitation.

From the topographical and aerial surveys a DTM model was built to delineate the whole catchment and its sub-catchments. The total area of 20.2 km^2 was divided into 23 sub-catchments of which 18 are located in the urbanized area and the other 5 are semi-rural, located in the upper part of the total area (Figure 7.1). Each sub-catchment is divided further in order to concentrate the outflows from the runoff at the nodes of the network; this increases the number of catchments being used by the hydrodynamic model. The total number of sub-catchments in the model is 136 (Figure 7.5). The selection of the rational formula coefficient "C" for each sub-catchment was done using Table 27, which is recommended by Venezuelan standard.

Table 27: Runoff coefficient "C" recommended by Venezuelan standard

| Land Use | Runoff Coefficient | | |
| | Mean terrain slope | | |
	S < 2%	2% < S < 7%	S > 7%
RESIDENTIAL			
Low density	0.45	0.50	0.55
Medium density	0.48	0.58	0.65
High density	0.65	0.75	0.85
COMMERCIAL	0.75	0.85	0.95
INDUSTRIAL			
Light	0.75	0.78	0.80
Heavy	0.80	0.88	0.95
RECREATIONAL			
Parks	0.25	0.30	0.35
Green areas	0.15	0.20	0.35

In order to build the drainage network of Cabudare it was necessary to model three main aspects of the drainage. The first is to model the stream (Tabure River), the second is the minor or below-ground pipe network, and the third is the major or above-ground street system.

For the river modeling it was necessary to do a survey in order to define several cross sections. In total 22 cross sections were surveyed along the river. Tabure River is a stream that carries water only during the rainy season; the rest of the year it is dry or carries only a small amount of water. When the stream enters the urban area of the catchment, several control structures are present (bridges, steps, culverts, etc,). These structures have to be incorporated separately into the hydrodynamic model. Part of the stream is a natural channel while other sections have been modified or canalized using concrete. Figure 7.6, shows a collage of pictures taken during the topographical survey. These pictures show that the structures interfere with the normal water flow, promoting the formation of distinct backwater curves upstream. The sediment accumulated upstream of the structures increases these backwater effects.

The Venezuelan standard allows for some inundation in streets and free areas like parks, parking lots, etc. Table 28 shows the water depth values permitted during a flooding event using design rainfall. These values are aimed at maintaining safe driving conditions on the roads. Additionally to the limits on water depths, the standard establishes a maximum horizontal extend allowed for flooding (Figure 7.7). As mentioned, these limits are aimed to maintaining safe driving conditions, however the roads can be used to convey water during extreme events.

Figure 7.5: Urban sub-catchments flowing to the pipe network system

Figure 7.6: Control structures and channel cover in Tabure River

In order to take into account these additional capacities of conveyance and storage of water in the above-ground system a hydrodynamic model is required. Inconsistent with the Venezuelan standard using steady state flow, hydrodynamic modeling of above and below ground drainage systems will allow a realistic evaluation of the damage costs.

Table 28: Maximum water depth allowed on roads

Road type	Max water depth in centimeters		
	Longitudinal flow	Transverse flow	On depressions
Arterial roads	6	0	6
Distributive roads	6	0	10
Local roads	6	5	15
Special roads	6	0	5
Railways	variable	0	N/A

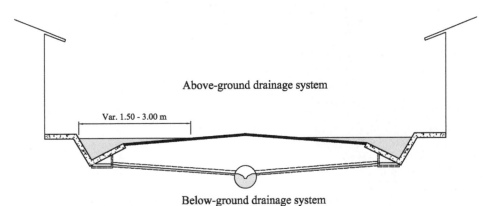

Figure 7.7: Cross section idealization for dual modeling and maximum allowed inundation in streets

The hydrodynamic model selected was MOUSE from Danish Hydraulic Institute (DHI). This enable the modeling of the river and the required above and below-ground drainage network. The connection between above and below-ground networks was done assuming that a manhole behaves like a weir which connects two nodes; one node is a manhole and the other is in the street. When the below-ground drainage system reaches its full capacity, the water level in manhole starts to rise until the ground level is reached; at that moment the weir start to convey the excess water to the street. This type of connection between the above and below-ground systems is able to transport water between different sub-catchments. If the water level is higher than the street curb, it overflows the curb inundating the urbanized areas.

The model was built using the sub-catchments, the pipe network and the street cross sections. Six models were instantiated, one for each return period (2.33, 5, 10, 25, 50 and 100 years). The whole drainage model of the Cabudare drainage system consists of 136 sub-catchments, 370 nodes and 370 links which represent the below-ground pipe network and above-ground street system. A layout of the drainage topology over a DTM is shown on Figure 7.5. A preliminary run for each return period was done in order to identify the problematic areas. From these initial runs it can be seen that even for a small return period of 2.33 years several areas are prone to flooding. These flooding areas were cross validated with information from the inhabitants of the areas during some recent events (2005-2007), and they match closely with the ones reported by the model. Figures 7.8, 7.9, and 7.10 shows the affected areas for several return periods. The maximum water depth is in the range of 1.00 to 1.60 meters.

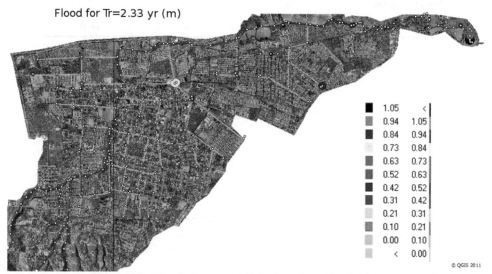

Figure 7.8: Flooded areas in Cabudare for a Tr=2.33 years

7.1.5 Urban Drainage Optimization for a Single Event

The Venezuelan standard recommends the use of a fixed return period for the design storm. It is dependent on the importance of the structures to be protected, the land use and the service life. Design return periods used to range from 2 to 25 years. In the case of below-ground and surface channel drainage for cities land use is the most important variable.; while for streets the importance of the road is the main variable to take into account. Tables 29 and 30 show the recommended values for this variables in Venezuela.

Flood for Tr=10 yr (m)

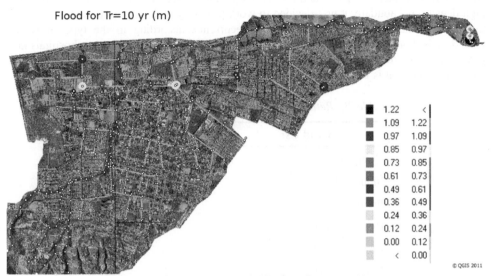

Figure 7.9: Flooded areas in Cabudare for a Tr=10 years

Flood for Tr=100 yr (m)

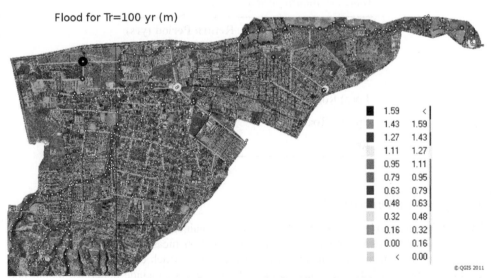

Figure 7.10: Flooded areas in Cabudare for a Tr=100 years

Because there exists a variable land use with a predominance of residential with medium to high density in Cabudare, it was decided to work primarily with a return period of 10 years that corresponds to the longest period of recurrence according to the type of roads. Cabudare is bounded by an arterial road ring and in the city center there are hospitals and government buildings (public areas) which also corresponds to the longest period of recurrence. Thus, it is justified the use of a 10 years return period.

Table 29: Return period according to land use

Land Use	Return Period (yrs)
Commercial	10
Industrial	10
Public	10
Residential (High density)	5
Residential (Medium density)	5
Residential (Low density)	2
Recreational	2
Green areas	1

Table 30: Return period according to road type

Roads Type	Return Period (yrs)
Arterial Roads	10
Distributive Roads	5
Local Roads	2
Special Roads	10
Railways	25

Investment Cost Function

The cost functions to be used follow the equations explained in Chapter 5. Pipe renewal and the use of storage (a wet pond) were the rehabilitation measures taken into account as investment costs. In the case of pipe renewal a catalog which contains the pipe cost per length of pipe for different diameters was built. These costs include materials, labour and machinery needed for pipe installation.

The storage location was an issue to address during field work. Four possible storage locations were identified; see Figure 7.11. However, only one was included in the optimization process. There is high pressure to build new urbanized lots in the area, this has made the land cost increase at a high rate. Also, after some meetings with local authorities (municipality engineer) and house builders, they expressed their intention to support the building of storage outside of the urbanized area, preferable in the limit of the urban polygon. In fact, for 2011 new areas have been urbanized, which were empty lots in 2007. For this reason storage ponds S2, S3 and S4 were discarded and only the pond S1 was considered in the optimization process. The S2 storage is inside the urban area; S23 and S4 are located down-stream and the will not be so effective in flood control; additionally they are located in the expansion polygon of the urban area.

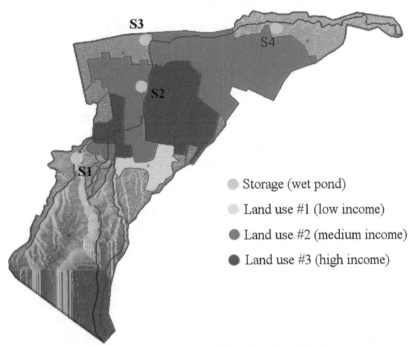

Figure 7.11: Storage location and land use for Cabudare city

The location of the storage S1 is upstream of the urbanized area. It is located in a lot which belongs to the local University, more specifically to the agriculture department. In principle, the use of a dry pond is better if the intention is to reduce flooding damage. However, the difficulty in finding a suitable area to store water temporarily, led to a negotiation with the University. The Cabudare region is semi-arid and the agricultural department can use store water for irrigation purposes. For this reason a wet pond will provide a combined use, namely to reduce flooding in the urbanized areas located down-stream and to use part of the stored water for the irrigation of crops in the experimental areas of the agriculture department. From the DTM of the area, the capacity curve of the storage was build, giving a maximum possible total volume of 58,000 m^3 for a dike 10 meters high.

The equation of the objective function for the investments costs is given by Eq. 80.

$$C_I = C_P + C_S = \sum_{i=1}^{np} \left[C(D_i) \cdot L_i + P_i \right] + \sum_{j=1}^{ns} 1680 \cdot V_j^{0.58} \tag{80}$$

where:

	C_I	is the total investment cost
	C_P	is the pipe investment costs
	C_S	is the storage investment cost
	i	is the index for pipe i^{th}
	i	is the index for storage j^{th}
	np	is the number of pipe to be changed
	ns	is the number of storage to be evaluated
	D_i	is the pipe diameter i^{th}
	$C(D_i)$	is the cost for the pipe i^{th}
	L_i	is the length of the pipe i^{th}
	V_j	is the volume of wet pond storage j^{th}
	P_i	is a penalty function for pipes not allowed pipe

Damage cost function

Damages due to flooding in the area were estimated as a function of the inundated area. For each node of the street network a curve that relates water depth-area-volume-cost was built. It was assumed that the maximum damages cost will be fifty percent of the total price of a house; this amount includes the house itself and the damage to equipment and belongings inside of the house. The maximum damage will occur when the water depth is two meters.

The area has several different types of construction and land use; and there is a predominance of residential. The total urbanized area was classified in three main residential areas depending of the socioeconomic class and type of urbanization (Figure 7.11). For instance, the land use #2 is for a socioeconomic class of a medium income, and a house price for this group is around BsF 600,000.00 or € 100.000 (2010 exchange rate). The average lot for this type of construction is about 400 m². This gives a cost per square meter of 120 €; The same analysis applied to low and high income houses gives 75 €/m² and 225 €/m² respectively. With these maximum costs, the relation between water depth damage discussed in Chapter 5 and the use of a GIS system (Grass) a curve for each node was built as shown in Figure 7.12. The final damage equation is given by Eq. 81.

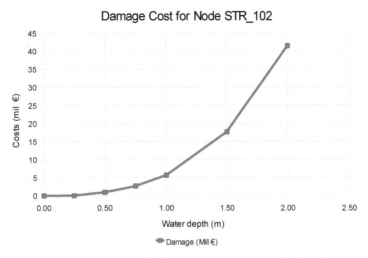

Figure 7.12: Water depth damages cost for node STR_102

$$C_D = \sum_{i=1}^{nd} \sum_{k=1}^{nlu} \left(fc_i^k \cdot C_{max}^k \cdot DEP_i \right) \tag{81}$$

where:

C_D is the total damage cost

fc_i^k is the value of damage factor at i^{th} node

DEP_i is the value of water depth in i^{th} node

C_{max}^k is the maximum damage cost for land use k^{th}

nd is the number of nodes

nlu is the number of land use

Optimization for a 10 years return period

Following the framework described in Chapter 3, the first step was to run the model using the current state of the network using the the rainfall for a 10 year return period. In this case the ratio Q/Qf was set to 1; the reason for this is that pipe inverts are close to ground level (1.2 -2.5 m) because it is a flat area and any small surcharge will flow above ground easily. Another reason is try and follow the Venezuelan standard which advises the use of the full capacity of the pipe networks. Once this run was done the algorithm was able to identify 81 pipes suitable for replacement.

This initial run was also used to get the maximum damage cost expected for a rainfall of return period 1:10. This also correspond to the "*do nothing case*". A second run was done assuming that the whole system will be upgraded using larger pipes from the catalog and maximum storage capacity; this result gives an idea whether there exists a solution that can produce zero damages. Obviously, this solution is not optimal but provides an upper limit on the investment. These two values: the maximum investment and damage cost are used to normalize the rest of the data.

The optimization was run 10 times. The population size of the GA was set to 24 individuals while the number of generations was set to 80 for a total number of function evaluations of 1920. The most important genetic parameters were set as follows: probability of crossover 0.90, probability of mutation 0.25, distribution index for crossover 15 and distribution index for mutation 20. An INTEL core duo processor running at 2.4 GHz was used. The total time expended for a single run was 18.3 hours.

The best Pareto set of ten is shown in Figure 7.13. The algorithm was able to find the "*do nothing*" solution which matches the initial run. Also it was capable of finding several solutions that produce zero damage, which are cheaper than the one assuming maximum investment. From the Pareto set it can seen that the maximum damage produced by a storm of 1:10 years is 3.34 millions euros. In order to protect the city from such damages an investment of 8.46 millions of euros has to be made, reducing the damage costs to zero.

Between the extreme values in the Pareto set, there are more than 20 values which provide compromise solutions between investment and damage costs. A gap can be identified in the damage costs between 0.86 and 2.17, in which the investment is almost the same. This corresponds to a change in a key pipe. Such a pipe is a large but controls the flooding upstream, and when it is changed a large reduction in the damage cost is obtained. Although a wet pond was used instead of dry pond and it was assumed to be at a normal level during the event, it was selected as part of the 20 solutions of the Pareto set. The storage volume is in the range from 40,000 to 50,000 m^3.

It looks as though in order to reduce the damage costs due to flooding, it is necessary to invest more (8.46 mil) than the benefit obtained (3.34 mil). If the damage cost is added to the investment cost the result is the total cost; in theory the ideal is to select a solution with the minimum total cost (Table 31). From a simplistic point of view the answer could be to do nothing; however, it has to be considered that during the life span of the system, several return periods will occur, so the accumulation of damages can justify taking some different actions.

Optimization using expected annual damage

The optimization using a single return period does not provide much information about damages for other return periods, and the accumulation of damages during a time frame. In order to address this issue, a method based in probabilities is needed. The Expected Annual Damages (EAD) method was used to calculated the expected damages (USACE, 1996). This method is based on computing the expected annual damage through the integration of a risk function.

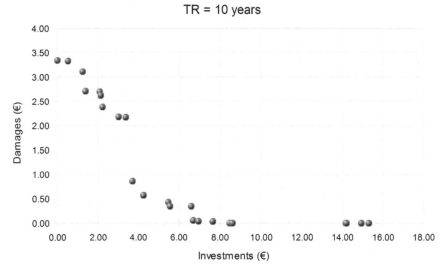

Figure 7.13: Best Pareto set for a 10 years return period for Cabudare city

Table 31: Pareto Set and total cost for Tr=10 years

Event of TR 10 years		
Investment	**Damage**	**Total cost**
0.00	3.34	3.34
0.52	3.33	3.85
1.24	3.11	4.35
1.39	2.71	4.09
2.08	2.70	4.77
2.13	2.62	4.76
2.21	2.38	4.59
3.02	2.18	5.20
3.37	2.17	5.54
3.70	0.86	4.56
4.23	0.57	4.80
4.25	0.57	4.82
5.46	0.43	5.89
5.55	0.34	5.90
6.59	0.34	6.93

Event of TR 10 years		
Investment	**Damage**	**Total cost**
6.68	0.05	6.74
6.94	0.04	6.98
7.64	0.03	7.67
8.46	0.00	8.46
8.59	0.00	8.59
14.15	0.00	14.15
14.19	0.00	14.19
14.92	0.00	14.92
15.28	0.00	15.28

The annual flooding risk has an exceedance probability (p) between 0 and 1, which is an indirect measure of the flood severity. This probability is the inverse of the return period ($Tr = 1/p$). The total expected annual damage is the damage that occurs for all possible flood probabilities as given by Eq. 82.

$$D_E = \int_0^1 D(p)\,dp \tag{82}$$

where: D_E is the expected annual damage cost

p is the probability ($1/Tr$)

$D(p)$ is the damage for given probability p

The EAD gives the damage that can be expected in a year. If the life span of the sewerage assets is estimated as N years, a uniform series can be defined and its present worth can be calculated. Applying this approach using Eq 81 and Eq 82 in discrete form the new damage functions can be rewritten as given by Eq. 83 .

$$D_E = \sum_{j=2}^{NTr} \left[\left(\sum_{i=1}^{nd} \sum_{k=1}^{nlu} (fc_i^k \cdot C_{max}^k \cdot DEP_i) \right)_{j-1} + \left(\sum_{i=1}^{nd} \sum_{k=1}^{nlu} (fc_i^k \cdot C_{max}^k \cdot DEP_i) \right)_j \right] \cdot \frac{\Delta p \cdot f_{PW} \cdot N}{2} \tag{83}$$

where:

D_E	is the expected annual damage cost
p	is the probability ($1/Tr$)
Δp	is the equation ($p_{j-1}-p_j$)

f_{PW} is present worth factor $f_{PW} = \dfrac{(1+r)^N - 1}{r(r+1)^N}$

NTr is the number of discrete return periods to use

N is the number of years of life span assets

r is the interest rate

The equation for computing EAD was added to the algorithm for the multi-objective optimization. Also a new feature was included in order to accelerate the algorithm convergence. The calculation of EAD requires the execution of the hydrodynamic model as many times as the number of return periods. It makes the determination of the Pareto set even more complicated and computationally demanding.

A new introduced feature, to improve the convergence speed, consists in "*inoculation of good genes*" into the initial population of the NSGAX. The former method was to let the population converge until all the solutions are formed, including the "*do nothing*" case and the maximum cost investment. It was explained above that these two values are known beforehand because the runs are needed for identification of the pipes to be replaced and for the normalization of the variables in decision space. Experts and practitioners can include solutions that they may want into the GA initial population. This allows the inclusion of expert knowledge to the optimization process, which is an integral part of the methodology described in Chapter 3.

The parameters of the GA program were the same as used for the 10 years return period rainfall, except the population size which was set to 12 and the number of generations which was set to 24 for a total number of function evaluations of 288. The other parameter related to the estimation of EAD were as follows:

- The number of return periods to evaluate was six: 2.33, 5, 10, 25, 50 and 100 years

- The rate of interest for the present worth was 2.22%

- The life span for the sewer network was estimated as 50 years

Five realizations were carried out, each complete run lasted 12.75 hours. The best result of the Pareto set is shown in Figure 7.14; Table 84 shows the results in tabular format. It can be seen once again that the Pareto set develops a concave shape with only 12 values. It can be noticed that the Pareto never reaches the investment axis. This means that there is no solution that reduces the expected damage to zero. This behavior has sense because there will always be a return period that exceeds the capacity of the drainage system, then the Pareto must be asymptotic to the "*x*" axis.

It can be observed that in the case "*do nothing*" the damage cost now is 35 million euros.

This value is 10.4 times larger than the value for a single return period optimization for 1:10 rainfall. The use of the accumulated damage costs during a life span of 50 years at a rate of 2.22% increases the value of the damages at around 35 millions of euros.

As for the single return period optimization for a 10 year of return period, a jump in damage costs for a investment of around 3.5 millions of euros can be observed. This is the same reasons as the single return period optimization; It corresponds to a long pipe which is changed and has a high impact on the flood damages.

If the total cost is computed by adding the damage and investments costs a total damage curve is obtained. A maximum and a minimum can be identified in such a curve. There is a maximum of 35.62 millions euros of total expected cost, see solution #2 on table 32. This solution is more expensive than the option "*do nothing*" which is 34.85 millions. If we look only at the Pareto, this solution has less damage cost but when combined with the investment cost, it becomes expensive.

There are three minimum values between 12 to 13 million Euros of total expected damage costs. These occur for investment costs between 4.6 to 8.6 million euros. It looks as though an optimal solution is in such a range and more specifically for the solution #4; this has a minimum value of 12.16 million euros for total expected cost for an investment of 4.6 millions euros. However, the other two minimum solutions have almost the same total expected cost despite higher investment costs. This means that decision makers have some almost equally ranked solutions with which to take decisions.

A comparison of the investment costs was done for the expected damage cost optimization and the single return period optimization. It can be seen that for total protection against a flood of 10 years return period 8.46 millions euros have to be invested. If this value is looked up on Table 32, it can be seen that it correspond to solution #6, which is not the total minimum (solution #4) but is very close. It is argued that the selection of the solution #6 is a good choice because it minimize the expected damage costs, while it also gives a total protection for the 10 years return period as expected in the Venezuelan standard.

Regarding computational cost, the addition of expert knowledge through the manipulation of genes in the NSGAX algorithm, allowed for a decrease in the total number of function evaluations and thus reduced the total execution time. The inclusion of two points only does not influence the radom search of the algorithm, allowing a good spread of solutions.

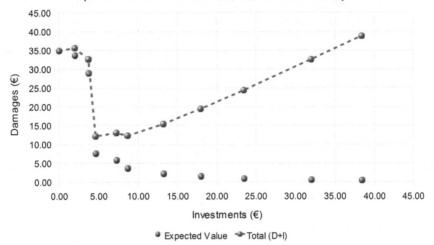

Figure 7.14: Pareto front for EAD and total expected cost (D+I) for Cabudare city

Table 32: Pareto front and total cost for Cabudare City

	Expected Value		
Sol #	**Investment**	**Damage**	**Total (D+I)**
1	0.00	34.85	34.85
2	2.02	33.60	35.62
3	3.74	28.88	32.62
4	4.60	7.56	12.16
5	7.25	5.78	13.03
6	8.66	3.63	12.29
7	13.18	2.21	15.39
8	17.95	1.51	19.46
9	23.48	0.90	24.38
10	32.00	0.53	32.53
11	38.37	0.42	38.79
12	46.55	0.30	46.85

7.2   Water Quality Application – Cali, Colombia

Cali is Colombia's third largest city. Founded in 1536 by "Sebastian de Belalcazar", it was a sleepy little mountain town until the sugar and coffee industries brought prosperity to the region. The city is located in Colombia's southwestern region, about 1000 meters above sea level. The urban area of the city is located in the Valley of the Cauca River. The region's climate is tropical with an average temperature of 25° C. The average precipitation is 1200 mm/year, but it can vary between 900mm/year in the valley to 2000 mm/year in the mountains.

According to the official census of 2005, the city has a population of 2.1 million inhabitants. The development of the city has generated a progressive growth mainly caused by the industrialization process that started in the 1940s. In 1938 the population was 102,000 inhabitants, in 1951 284,000 inhabitants, in 1973 918,000 inhabitants, in 1993 1.6 million inhabitants, and currently it is the third largest city in the country with 2.1 million inhabitants. Some characteristics that explain the rapid growth of the city are the economic development by the agro-industry, the location of Buenaventura, the biggest and main port of Colombia, just two hours away by car; and migration processes caused by internal conflicts in the country.

The development of the city has been at the expense of a high environmental cost, especially in terms of the disappearance of some of the rivers, wetlands and lagoons that regulated the hydrogeology in the Valley. In this sense there has been a lack of planning and integrated vision to develop an environmental friendly urban area that can take advantage of the resources available in favor of an improvement of the citizen's standards of living. The environmental problems and issues can be summarized as follows:

- The water quality in the Cauca river has been progressively deteriorating. This situation is critical since this is the main source of water supply for the city and the region;

- There are discharges of wastewater from the city to the river with high load of sediment and pollutants;

- The decisions for the control and management of the river are taken based on unreliable information;

- The development of the sewer system of the city has been done without integrated planning.

The present work explores the inclusion of additional objectives in the drainage network rehabilitation process. The purpose is to find the optimal sewer system rehabilitation strategies by using three objective functions, namely, rehabilitation cost, flood damage cost and pollution cost caused by sewer overflows. This time the Storm Water Management Model (SWMM version 5.0) is used as computation engine to simulate hydrological, hydraulics and water quality routing and processes in the system instead of DHI-MOUSE. In order to link the genetic algorithm with the computation engine SWMM 5.0 two intermediate link routines were written, following the schematic methodology presented in

Chapter 3. The first routine runs the original model in SWMM 5.0 and computes the magnitude of the flooding, the surcharged pipes, and the initial values of the variables needed for the objective functions. The second routine directly links NSGAX and SWMM 5.0 by translating the scenarios generated by NSGAX to SWMM input file, and computing the value of the three objective functions that are passed to the GAs for evaluation and generation of further scenarios.

The following steps are used to find the optimal alternative of rehabilitation works requirements:

- Initial simulation of the existing drainage network.

- Performance evaluation against given standards (constraints).

- Calculation of objective functions.

- Identification of the new drainage network set up (i.e., new rehabilitation strategy).

- Simulation of the new drainage network.

7.2.1   Objective Functions

Overflow caused pollution

A concentration of total suspended solids is assigned for different nodes as the dry weather flow and the event mean concentration method is used to describe the build up of pollutants in the catchments. Once the model is run in SWMM 5.0 the output reports the total amount (mass) of pollutants that are generated from the dry weather and wet weather flows.

If sufficient conveyance capacity exists within the network, then the sum of these two amounts will represent the total mass of pollutants that will end up in the recipient water body. In the case of flooding, the model accounts for the overland surcharge and in the same way it accounts for the total mass of pollutants that is contained within the corresponding mass of flood water.

To represent this situation and take into account the possible Best Management Practices (BMPs) for urban drainage, Figure 7.15 shows a conceptual model that illustrates a mass balance in the system. BMPs are referred to as measures that can improve the water quality and alleviate the discharge that goes to the receiving water body. Some of the measures include storage or detention basin with a certain level of treatment, infiltration, bio-filtration, etc.

The amount of water that is stored in this way is discounted from the flood damage objective function. Additionally, the stored water can have a certain level of treatment or removal of the pollutants. This is also represented as a fraction of removal and it is yet another variable used in the optimization process.

The objective function is formulated as a penalty function for water pollution. For this study the following amounts are assumed for the penalty function:

- Maxallow : 5000 (maximum amount of pollutant that can be discharged to the river without affecting the environment).

- C_1 : Cost associated to the discharge below Maxallow (assumed as 100)

- C_2: Cost associated to the discharge above Maxallow (assumed as 10000)

- C_3: Assumed cost penalty for having polluted water on the streets (10000)

- C_4: Assumed cost to discharge waste-water from storage with not treatment (100)

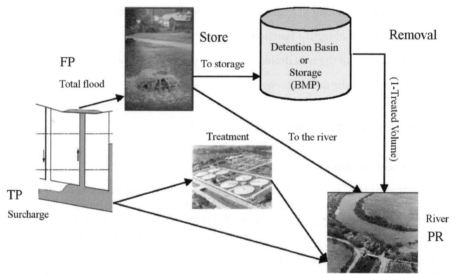

*Figure 7.15: Conceptual model to account for storage and treatment of flooding. **TP** = Total mass of pollutants (dry weather) + Pollutants (wet weather); **FP** = Mass of pollutants leaving the system by flooding; **PR** = Mass of pollutants that reaches the outfall in the river; **Store** = Volume of water to store; **Removal** = Fraction of pollutants that can be removed (efficiency)*

The objective function for the overflow generated pollution can be written as:

$$WPO = C_1 \cdot Maxallow + C_2 \cdot (PR - Maxallow) + C_3 \cdot FP \cdot (1 - Store) \\ + C_4 \cdot FP \cdot store \cdot (1 - Removal) \tag{84}$$

where:

WPO	Water Pollution Objective
TP	Total mass of pollutants (dry weather) + Pollutants (wet weather)

FP	Mass of pollutants leaving the system by flooding
PR	Mass of pollutants that reaches the outfall in the river
Store	Volume of water to store
Removal	Fraction of pollutants that can be removed (efficiency).

Flood damage function

The flood damage is computed based on the flood magnitude calculated at each node. This follows a similar approach as discussed by Tanaka and Tarano (2000), Vojinovic et al. (2006) and Barreto et al. (2006). The flood damage function is as follows:

$$FD(S) = \sum_{i=1}^{n} \beta \cdot \left[e^{\frac{(S_i - S_{allow})}{1000}} - 1 \right] \qquad (85)$$

where:

FD(S)	is the flood damage function of volume
S_i	is the flood volume at each node (ha-mm)
S_{allow}	is the maximum allowable flood volume (assumed 1 ha-mm)
n	is the number of nodes analyzed in the network
β	is a penalty factor (100.000 in this case), this value would depend on the value of property for instance

To account for the effect of the allowed storage volume on the total flooding, an equivalent fraction is deducted from the flood damage function during the evaluation of the above objective function in the optimization loop. The flood damage function can be expressed as:

$$flood\ damages = \left(1 - \frac{vol}{maxvol}\right) \cdot FD(S) \qquad (86)$$

where:

vol	is the storage volume of a particular solution
maxvol	is the maximum allowable volume of storage

Rehabilitation Cost

The total cost is a combination of actual cost and operational cost when the whole design cost needs to be considered. For example the cost of storage and waste-water treatment and the operation of the pumps will need to be considered. The infrastructural cost objective function includes the cost of pipe replacement, the cost of the storage of a fraction of the flood volume and the level of treatment of the stored water. The rehabilitation cost function can be expressed as:

$$RC = C_v \cdot S_{vol} + C_T \cdot T_{vol} + C_{WWTP} \cdot T_{WWTP} + \sum_{i=1}^{np} C_p(D_i) \cdot L_i \qquad (87)$$

where:

RC	Rehabilitation Costs
S_{vol}	is storage volume of flood in percentage
T_{vol}	is the fraction of store volume that receives treatment
T_{WWTP}	is the amount of pollutant treated at the treatment plant
n	is the number of pipes in the network
D_i	is the diameter of the pipe (i)
L_i	is the length of the pipe (i)
C_p	is the cost per unit length of the pipe (i)
C_v	is the cost of storage \$/m3
C_T	is the cost of the treatment system, it will depend on the BMP or treatment technology
C_{WWTP}	is the operational cost to remove the pollutant (assumed as 1.2 \$/Kg)

All results obtained from the objective function calculations are normalized within the interval [0,1]. Normalization typically provides an advantage for the graphical evaluation of the Pareto and the solutions.

An experimental application based on a case study in Cali was carried out in order to test the approach described in previous sections. The base model was built with SWMM 5.0; the model incorporates a node with waste-water treatment in a simplified way (the concentration of the pollutant that is removed was assumed to be 60% efficient). The receiving water in the river is also incorporated in a simplistic way to integrate the system and to facilitate an integrated analysis. Figure 7.16 shows the schematization of the network.

A typical design rainfall event for the area was used (i.e., 1:5 year ARI event). The initial run displayed a result with several locations being flooded. These are the locations which were used within the optimization process to adjust diameters of those pipes which are the cause of flooding and to minimize the flood impacts.

The probability of crossover was set as 0.9 and the probability of mutation was set as 1/*nreal*, where *nreal* is the number of real variables (Barreto et al, 2006). The total number of variables used in this study was 74. Out of the 74 variables, 72 correspond to the number of pipes in the model; the diameter is used as a variable and each pipe has a value in a catalog with 20 commercially available sizes. The last two variables are kept for storage and the removal of the pollutant. The idea of the remaining two variables is that at the end of the optimization process there will be more solutions to solve the problem of flooding and to reduce the concentration of a pollutant in the system which overflows into the river. In this way each specific solution has an estimate of the size of the storage required and the degree of removal of the pollutant which gives an indication of the BMP to be used.

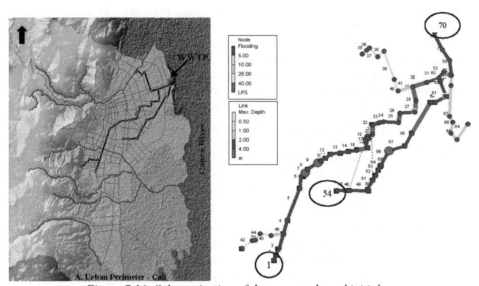

Figure 7.16: Schematization of the case study and initial run.

Results

Several trial runs were done with different configurations of the population size, the number of generations and a random seed; some of these runs took 24 hours on a Pentium 4 Centrino duo processor and 2 Gb of RAM . The trial runs indicate that with a smaller number of combinations, results are likely to be in the same range. The results presented consist in two runs that were done to perform the analysis with the same set of parameters. The test was done with a population of 100 individuals and 70 generations for a total of 7,000 functions evaluations, the same value of the seed was used (0.494) and the experiment lasted for 268 minutes. In the first experiment the maximum inflow of the link connected the treatment node was not limited. In the second experiment the maximum inflow to the treatment node was set to a maximum value of 2.7 m^3/s.

Figure 7.17 shows the results of the optimization process. Only the non-dominated points are plotted, the graphs are organized in pairs per objective function so that they can be observed more easily. During the optimization process 100 solutions were found in the best population kept by the NSGAX.

From the results of the optimization process presented in Figure 7.17 it can be observed that there are solutions that solve the problem of flooding and minimize pollution in the river simultaneously. It can also be observed that there is a certain level of infrastructure cost where the pollution function is at a minimum and which tends to be at a constant value thereafter, while for the same value of infrastructure cost the flooding is reduced to zero. Therefore, the NSGAX gives a set of solutions where the water pollution is at a minimum and remains constant, while extra infrastructure cost is added in solving the flooding problem.

An observation of the solutions of the hydrodynamic model (SWMM 5.0) shows that there are solutions that can solve the flooding problems and minimize the pollutant in the river. The system can achieve this by retaining the solutions where the diameter of the pipe connecting the WWTP is larger and it attempts to pass the maximum possible amount of water to the treatment node. In reality, most of the time allowing such an overload to the WWTP is not advisable and not allowed in many cases since it has a negative effect on the treatment process and the overall performance of the WWTP.

In the second experiment the maximum inflow to the treatment node was set to 2.7 m^3/s. The parameters to run the NSGAX were the same as for the previous case. Figure 7.18 shows the results of the optimization process. Only the non-dominated points are plotted; the graphs are organized in pairs per objective function so that they can be observed more easily. During the optimization process 100 solutions were found and only the best population was retained by the NSGAX.

Figure 7.18 should be studied carefully since the trends of the variables do not follow the same as in the previous case. An important observation is that the objective function for the pollution caused by the overflow contains a certain threshold or value of the maximum amount of water that can be stored and treated outside the system. That is the reason why in the plot corresponding to objective flooding damage and objective function water pollution there is an inflexion point that corresponds to the maximum storage capacity that is set up at the beginning of the exercise.

Due to the constraint imposed on the pipe that transports waste-water to the treatment node, not only the magnitude of the flood was found to be higher when compared to the previous case but also the amount of pollutant that gets into the river was found to be higher in comparison with the previous case. This certainly has an effect on the number of solutions that can solve the problem of flooding. For example, there were 12 solutions with zero flooding identified in the first experiment, whereas, in the second experiment there were only 2 solutions. Nevertheless, this experiment also shows that there are a number of combinations of the storage capacity and the removal of the pollutant that helps diminish the flooding and the pollution that reaches the river. A group of 17 solutions were found that combine the principles formulated for the objective function of the pollution caused by the overflow. This shows that the initial idea of introducing two additional variables for storage and removal of pollutant in the optimization process can be feasible. In fact, if both variables are considered together they could guide the selection of BMPs to improve the urban runoff water quality. Table 33 shows the denormalized range values for the Pareto sets for each case that was evaluated.

Table 33: Range values for the objectives function

Objective Function	Case 1		Case 2	
	Min Value	Max Value	Min Value	Max Value
Investment Costs ($)	573,000,000	582,000,000	571,000,000	599,000,000
Flood Damages (m³)	0.00	14848.00	0.00	24831.00
Water Pollution (kg)	25938.00	31821.00	28966.00	34627.00

7.3 Summary and Conclusions

In developing countries, the introduction of model-based multi-objective optimization may not be easy. The approach is computationally intensive and in many cases there could be a problem of lack of data. However, in this chapter we demonstrate that the presented approach can be valuable also in developing countries. Two two real studies case are considered: one case study was Cabudare in Venezuela and the other in Cali in Colombia, and innovative improvements to the approach were tested.

In Cabudare case study, an approach using expected damage cost was applied. It allowed to find three equally economic solutions in the Pareto set. A comparison with the recommended approach by Venezuelan standard shows that these three solutions allow to elimiate the flood damage completely for a return period of 1:10 rainfall. It also provides a set of several solutions in which there are a trade off between investment and damage. The NSGAX algorithm was improved allowing the inclusion of expert knowledge through the injections of potentially good solutions (good genes) into the initial population. This new approach proved to be efficient for computationally demanding problems.

In Cali study case the environmental objective function was considered. The multi-tier frame was coupled with SWMM 5.0 to compute flooding and water quality. A diversion was included to divert water from the network to a temporal storage for flood reduction and catching the first flush. Despite of the simplification on the storage model, it can be seen the trade off between the investment cost and flood damage, between investment cost and water quality and the correlation between flood damages and water quality improvement.

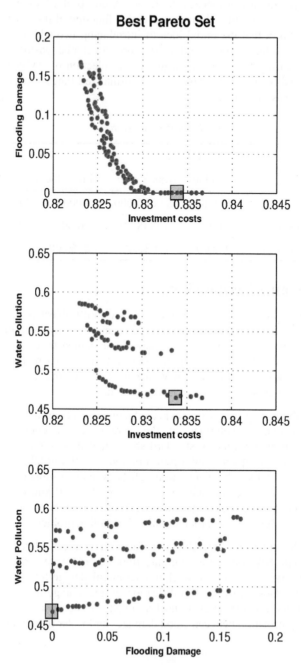

Figure 7.17: Normalized Pareto set for the case without limiting the flow into the WWTP. The grey squares show the solution with minimal pollution

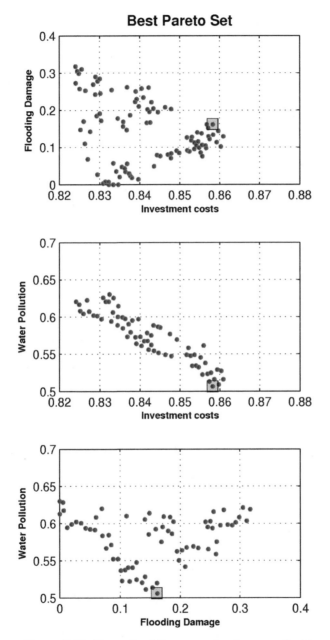

*Figure 7.18: Normalized Pareto set for the case limiting
the flow into the WWTP to 2.7 m³/s. The gray squares
show the solution with minimal pollution*

8 Conclusions and Recommendations

The objective of developing and testing a framework for the rehabilitation of urban drainage networks has been achieved. A prototype of the multi-tier approach was applied to several case studies in developing countries, demonstrating that it is feasible and easy to apply. Parallel computing and other methodologies make the approach attractive to practitioners, and have been included in the framework. The main findings, limitations and recommendations of the research are given in this Chapter.

8.1 Conclusions

- Data presented in Chapter 1 about population growth and urbanization makes it evident that the increasing urbanization around the world increases the demand for drainage services. The negative effects of urbanization impact more heavily on developing countries, where there is a lack of poorly designed master drainage plans for their cities, and legislation is either not applied or ignored by city administrations. Whereas in Europe the problems of urban drainage regarding quantity and quality are not such a major concern, it is certainly a crucial concern of cities in South-American countries.

- There is a need for tools that can help planners and decision-makers develop and implement effectively drainage plans that allow for negotiations with stakeholders to determine fund allocations. In the last decade, tools like CARE-S (2001) and Hydroplan (2005) have been developed in Europe, but they have not been widely used due to their complexity. Application of such tools in developing countries becomes even more difficult due to the lack of basic data.

- A multi-tier framework has been developed to carry out the rehabilitation of urban drainage networks. It is flexible and can integrate several objectives and levels of performance or costs, allowing their optimization in a multi-criteria framework. The block structure of the framework makes it flexible, permitting the inclusion of other external tools for hydrodynamic modeling, damage cost estimation, and derivation of investment costs. Also, the block structure of the framework provides scalability: it can start with simple models, depending on data availability, and increase their sophistication to the required complexity for the city. This scalability feature is an advantage in developing countries, where data is often scarce or unreliable.

- As explained in Chapter 2 and Chapter 3, the hydraulic model and the way that it is instantiated play a major role in any rehabilitation plan. Above and below ground simulation has been successfully used as a method for better evaluation of hydraulic performance indicators. This approach allows for the transfer of water between sub-catchments, giving a better hydraulic simulation if only the minor system (the below ground pipe system) is simulated; in this way the function for damages costs represents a more realistic damage scenario. In cases such as Cabudare city in Venezuela, where legislation allows for the streets to be partially inundated, the load on the below-ground pipe system is reduced during the rehabilitation optimization.

- OPTRESS, the tool for the OPTimal REhabilitation of Sewer Systems, has been developed and tested. It makes it possible to identify the optimum diameters of pipes through the use of the ACCO algorithm, and takes into account the poor hydraulic performance of pipes following the methods presented by the Hydroplan project. This software follows a single-objective search optimization approach, but it is able to come up with a set of sub-optimal feasible solutions which, although they are sub-optimal, can be considered by decision-makers or practitioners for other purposes. The tool developed to identify low performance in pipes was used in all case studies of this thesis and shows good performance in each case..

- In Chapter 4 several methods for optimizing using a multi-criteria approach were discussed as needed for an urban drainage rehabilitation plan. However, there have been few applications to real drainage system networks. Genetic algorithms are based on populations and generations and can move toward optimal solutions. They are suitable for scenario generation in the multi-tier rehabilitation framework; these scenarios are evaluated from hydraulic, environmental and structural points of view. The decision about what is the best method was addressed in this thesis. A comparison of two multi-objective optimization algorithms, based on GAs, has been made. NSGA-II and ε-MOEA were used as the optimization engines. The concept of ε-dominance in ε-MOEA gives a more robust Pareto, where the solutions are more evenly spread, than NSGA-II. However, NSGA-II provides a Pareto with more optimal solutions or scenarios; it is even capable of finding extreme values for each objective, including the null or "*do nothing*" " scenario. In the end, more solutions for the Pareto set are desirable if unfeasible scenarios are to be discarded; this implies that NSGA-II would be typically a better algorithm to use.

- When an optimization process is carried out using a GA approach, the norm is to set a population size and a number of generations; this effectively fixes beforehand the number of evaluations of the objective functions during the optimization procedure . It has a drawback when the problem being solved is highly computational demanding, as it is in urban drainage rehabilitation where the hydraulic evaluation is done using hydrodynamic models. In such cases computation time becomes an issue: if more function evaluations are done than are needed time is wasted, but on the other hand, if fewer function evaluations are done there is a risk of non-convergence to an optimal Pareto set. This thesis has solved this problem using two well recognized metric indicators for comparing

Pareto sets: Hyper-volume and ε-Indicator. Methods using both indicators show convergence, and can be used as a stopping criterion during the optimization, avoiding unnecessary further evaluations of the objective functions. However, Hyper-volume is more stable than ε-Indicator and is recommended as the better stopping criterion.

- NSGAX, a prototype tool implementing the NSGA-II algorithm, has been developed. The design of this tool allows for integrating hydraulic modeling software with cost estimation tools and multi-objective optimization algorithm for minimizing any numerical objective and particularly it has been used to minimize damages costs, investment costs, river pollution and anxiety in people. NSGAX follows the multi-tier framework presented in Chapter 3. The implementation was done using a dynamic link library for Windows with its flexibility to be incorporated in any programming language. It was also ported and successfully used with MS-Excel for rainfall-runoff multi-objective calibration of the HBV model.

- There are several kinds of physical measures that can be implemented for the rehabilitation of an urban drainage network. Measures of rehabilitation in terms of managing rainfall volume and conveyance are the two main approaches; volume control is preferred over conveyance because it leads to a significant reduction in peak flows and also because of improvements in water quality. Storage tanks, wet and dry ponds, overflow diversions, and pipe replacement have also been successfully applied to solve rehabilitation problems in the developed framework. The way that they were introduced, using the same hydrodynamic model, facilitates the optimization process and avoids the use of additional external tools for such a task. The external tool has only to compute the rehabilitation investment costs from the pipe or storage dimensions. In the case of storage tanks and wet or dry ponds, the methodology behaves in rational way by selecting the cheapest combination of pipe diameters and sizes of the ponds.

- Tangible and intangible costs were incorporated in the presented methodological framework. This work tests a couple of damage cost functions due to flooding. One such cost function based on an empirical exponential function was implemented for flood damage; real damage costs are usually found to be exponential with depth, but this type of function needs to be calibrated in order to obtain reasonable results. The other cost function for damage evaluation was based on a depth-area-cost curve that is constructed for each node offline, using the water depth at the node, land use and the digital elevation model combined with a GIS. According to the land use, a maximum damage cost is estimated, and depending on the flooding depth, a percentage of the total cost is calculated. This last approach gives better and more realistic results on real applications than the exponential equation, as is shown in the Cabudare case study.

- Incorporation of intangible cost has also been evaluated. One of most difficult tasks is the estimation of intangible damages. An equation for the anxiety of population living in a flooding area was developed based on the Lekuthai and Vongvisessomjai (2001) equation. This new equation, in addition to the flooding depth, introduces factors for mutual support, the percentage of vulnerable persons, past experience of flooding and an income factor. Anxiety is computed by applying the equation to each node, deducing a numerical value and adding the results. In this way anxiety was incorporated into the framework developed in this thesis.

- Highly computationally demanding simulations of water related problems are frequently encountered. Problems related with real time control (RTC), early warning system (EWS), uncertainty analysis and optimization are good examples of such demanding problems. Optimization of an urban drainage system comes under the category of such problems, especially if hydrodynamic models, whether 1D or 2D, have to be used inside the optimization loops. In this case the time required for running the whole process may not be feasible for practitioners. Two main approaches are used in this thesis to solve such kind of problems: one is to increase computational power and the other is to decrease the number of the function evaluations through an improvement in the algorithms. A multi-objective optimization method based on single search (MOSS) that is suitable for computationally demanding problems has been proposed and developed in this thesis. The method, which is based on weighting of the objectives and solving a minimization problem using a single search optimization was tested with four benchmark functions for multi-objective optimization. The single optimization was done three times by changing the weights and it outperformed NSGA-II on three of the benchmark function in terms of the number of function evaluations; for the fourth benchmark equation NSGA-II was better. This proposed approach looks promising for this kind of problem. The algorithm was improved through introducing an anisotropic elliptical search.

- The increase in computational power with respect to water management-related issues is achieved mainly in two ways: the use of faster computers, namely multi-core computers and clusters (supercomputers are indeed rarely used). Personal computers have reached a limit in speed with current technology, so it is necessary to introduce other technologies in order to have faster computers. Multi-core processors are becoming the mainstream technology nowadays. However, in order to use their power it is necessary exploit parallelism in the development of appropriated algorithms. (On of the problems is that all cores use the same file system so that some files may be locked being opened by one core, whereas another core may initiate write operations at the same time.) Parallelization of the multi-tier approach with the NSGAX algorithm has been implemented using the PVM library. This library was selected because it provides a stable tool that can be used on multi-core PCs and clusters of PCs as well. Two test applications were carried out on the parallel platform: a small pipe network and larger real case at Belo Horizonte in Brazil. It was demonstrated that there is an improvement when the framework uses the parallelization approach; for the small pipe network the improvement was 62.5% of the computational time, while for the Belo Horizonte case it was about 80%. The comparison was also done using the efficiency regarding the number of the processors; it was demonstrated that the use of more processors does not imply a linear gain in eficiency. If energy cost is considered the optimal number of processors has to be estimated. The results show that the larger the problem the more efficient is the parallelization. A cluster was set up over a non-specialized LAN and no optimization of the network components was done. The libraries and operating system of the master computer were open source and freely available. This makes the approach low-coast and suitable for developing countries.

- Expert knowledge was included in the framework at two stages of the rehabilitation procedure. The first stage for knowledge inclusion is during the selection of pipes suitable for rehabilitation where we assumed that practitioners may want to take part in this process.. The second stage for expert knowledge inclusion is during the optimization process: gene manipulation in the GA is done by introducing possible scenarios for the solution as selected by an expert. These solutions represent a set of good genes that will improve the search for new scenarios or solutions. The NSGAX algorithm was modified in order to include specific scenarios given by an expert; these scenarios are included in the initial population of the GA; it is like inoculating the parents with "good" genetic code. The solution "*do nothing*" and the maximum value of investment for rehabilitation problems are the most obvious scenarios to include. Other solutions identified by a drainage expert or evident to practitioners can also be included. Their application to the Cabudare – Venezuela case study shows how using this approach, the population size and the number of generation are reduced, leading to the reduction in the number of function calls

- From the results obtained from the Cabudare case study, it can be concluded that cost evaluation functions for damage in urban drainage systems using a specific return period should be adopted in conjunction with risk and using the expected value. The use of a single return period gives a snapshot of one event and the damages that it can produce; however, these damages are higher if a project horizon is taken into account. From an economic point of view, the selection of a single return period does not give much information about the cumulative damages. The correct selection of the design return period has to be done from a hydro-economic approach. It means that the urban system has to be designed or rehabilitated for a return period which minimizes the total cost over the life span of the systems. In Venezuela for instance, legislation gives as rule of thumb the use of a return period between 2 and 10 years, but the final return period has to be selected by the expert or practitioner, which is a subjective task. The application to Cabudare city has demonstrated that the system has to be rehabilitated for a return period of ten years, which minimizes the expected total cost. The use of an expected value for the optimization produces a more robust solution than a single design rainfall event.

- Real case applications have been carried out using the multi-tier rehabilitation framework, with application to three cities in developing countries. The literature review has shown that there are few applications of multi-objective optimization for drainage rehabilitation coupled with hydrodynamic models to compute damages inside the optimization. To the author's knowledge, before the introduction of multi-objective optimization by the author on 2006 there was only one application in drainage design. Applications to real cities were not addressed previously because the optimization process is computationally demanding, and very time consuming. In developing countries application of complex approaches are rejected from the beginning due to the lack of data. The optimization approach was successfully applied to cases in developing countries due to the scalability which allows the use of simplified or complex modeling tools. In developing countries, the use of clusters is not widespread but it offers a cheap solution to increase computational power. During the project the parallel processing software that was developed was set up on a memory stick, which only requires running an installation batch file, making it suitable for any office or small laboratory. The framework was also tested using both commercial and free software. MOUSE and SWMM were tested and successfully implemented in two cases in Latin-America. The parallel code was set up using PVM and Cygwin, which are open source libraries; thus making the code accessible for any developing country at no cost.

- Some drawbacks were found during hydrodynamic model testing. During several tests, SWMM was found to perform faster than MOUSE. A possible explanation is that MOUSE runs using a server license: checking the license over the network takes more time than when it is not checked. For a single run the difference in run time is imperceptible, but when thousands of runs are performed it can increase the computational time substantially. Another problem was the use of constraints using penalty functions. For example, constraining pipe diameters to be increasing downstream during the optimization process, does not generate good results. Such an approach reduces the feasible domain resulting in quite poor performance of the optimization algorithm; it was better to use post-processing tools for these kinds of constraint.

8.2 Limitations

- The framework as it was initially conceived, does not take into account investments during the life span of the drainage network. If is true that the use of several return periods for determining the expected value during the service life of drainage network helps us to consider the total amount to be faced in the rehabilitation project; the tool is not able to allocate this amount in an optimal way if the project is planned by phases; in other words, in its present form the framework is not a dynamic tool for determining investment (however, it can be modified to be such).

- The structural condition of the pipes, in the developed framework, is taken into account with the age of the pipe. A structural assessment for the pipes in the drainage system is required in order to develop an effective rehabilitation plan for the system. Although this can be introduced using the expert knowledge, it is better to introduce a pipe failure model, or to develop an automatic methodology using CCTV images to help in their selection. Also, other approaches can be included to find pipes in a drainage network that perform under the average. In this thesis the Q/Qf ratio was used successfully; however, a gradient-based approach as showed on Chapter 2 could be better.

• Optimization using MOSS has shown its efficiency, but still it was not included in the framework prototype. While this tool is able to find some values on the Pareto set quickly, not all the solutions can be found. However, MOSS algorithms has to be developed and tested further since its strength is for computationally demanding problems: the speed with which (limited number) of Pareto solutions can be found by the method can be very useful for practitioners.

• It is a fact that PVM is very easy to implement for a test case; however, since the algorithm was implemented in 2006 the tools for parallelism has been improved significantly and the trend is migration to MPI protocols. In the case of multi-core computers parallelization is done through a compiler using OPENMP. MPI looks more suitable because it also works on a cluster. Even though the protocol was developed for multi-core computers it can be used for code parallelization on PC clusters, while OPENMP can only be applied to muti-core computers.

• One of the major difficulties of this research has been testing of the framework for the cases in developing countries. In such countries there is often little information available in general. This makes it rather difficult to set up models and calibrate them. Cost function estimation is also difficult due to lack of information. For the case study in Venezuela several meetings were held with the municipality and water central water administration; without their willingness to collaborate and to go beyond the traditional approaches it was difficult to get the available information. Sometimes it was easier to find the information directly from contractors and the private companies in charge of the service maintenance.

• Simplified descriptors of hydrological events were used due to the lack of data on the real events. These latter happen to exist but it was impossible to obtain the data from the water authority. Instead, a set of synthetic rainfalls were used; the Chicago method was used to distribute the rainfall depth in time from the IDF curves. When the monitored data is made available it can be easily fed into the existing models.

• Another limitation is related to flooding along channels. When the main drainage system uses channels instead of pipes, flooding occurs at the extreme ends of the channel (nodes) and not along it if 1D modeling is used. This can be partially overcome using short length of channels or using a 2D model.

• Change in objective functions still require changes in the source code that needs to be recompiled; their implementation using external libraries would be a better approach.

8.3 Recommendations

• Further research has to be focused on consolidating the multi-tier framework, and transforming it into a software tool for DSS within the urban drainage rehabilitation process. Data was managed using text files, so a friendlier front-end GUI would be recommended to develop. A database and GIS for input and output data are needed since they will facilitate data management when large drainage networks are to be rehabilitated.

- Development of the better visualizations tools is recommended. One of the main issues to address, when multi-objective is used, is the visualization of data and results. When two objectives are employed, there is no major concern. However if more than two objectives are employed it becomes very difficult to interpret the Pareto front, as can be seen in the Cali case study. Virtual reality has been tested for visualization in three- dimensional space (Madetoja et al. 2008) but it is still not tested in the field of water resources as a DSS tool. When the Pareto frontier is generated for more than three objectives visualization becomes more difficult. Other approaches have to be developed; self organizing maps could be explored as a visualization tool for multi-criteria analysis.

- Instead of considering costs associated with a generated rehabilitation solution, it would be advisable to consider the total use of energy as a result of implementing such a solution. It is also important if one wants to compare similar projects to be implemented in developed and developing countries. Comparing monetary costs between countries is very difficult. Research findings, in which money is the main unit, cannot be easily transferred to other parts of world. Too many factors affect the real price of the money; in other words, there is strong subjectivity in this approach. During the 1973 oil crisis caused by an Arab oil embargo, the US started research to optimize energy consumption. One of the approaches proposed by the researchers was to optimize and evaluate all the processes in terms of energy. However, after the crisis was over this approach was not developed further. Any process, activity or material can be expressed in terms of energy; it is an absolute scale which is independent of location and income. In this way, for instance, damage costs can be compared between regions. Also, it will be an indicator of economy or energy savings.

- Fuzzy objectives for intangible cost can be incorporated inside the optimization process. In the test of an intangible cost, the anxiety objective was measured as a numerical variable. However, not all such variables can be measured this way. Some variables are qualitative and have to be treated as non-numeric (class) variables. Final defuzzification can be used to equalize the scales of non-numeric and numeric data.

- The incorporation of dual modeling, 1D-2D, for the above and below ground drainage systems has to be investigated and evaluated. Even though dual modeling was implemented in this thesis, it was done using a 1D-1D hydrodynamic model. This approach does not represent very well the flooding over the urban surface, so the estimation of damages is inevitably simplified. Dual modeling using 1D-2D hydrodynamic model coupling was envisioned at an early stage when this research was started. For well established networks with a realistic configuration such combination can be implemented without too much trouble. However, when a combined model has to be run for various sets of parameters (pipe diameters and storages) inside an optimization loop hundreds of times, we found that there are combinations of parameters that make the hydrodynamic computations unstable and produce erroneous results which and a modeller has no possibility to detect timely. Unfortunately the available time did not allow for running more experiments. Fortunately, models are becoming more stable, warranting optimization using 1D-2D models run in unattended mode a real possibility. The implementation of parallel computing using clusters makes this kind of modeling also more accessible.

- Due to the nonexistence of data for calibration in real case applications, it is necessary to implement uncertainty analysis. To some extent the uncertainty is implicitly taken into account by the ε-MOEA algorithm used in this study, however full-fledged analysis was not assumed to be in the scope of this thesis. However, the task is facilitated by the parallel approach of the multi-tier framework. An uncertainty analysis must be carried out after the Pareto is obtained. Each scenario in the Pareto set must be submitted to the analysis of uncertainty; in particular to the parameters of the rainfall-runoff and hydraulic models. Uncertainty in the cost estimation is yet another aspect to be analyzed. Overall, the uncertainty analysis will provide an idea of how robust the solutions contained in the Pareto set are.

References

Abebe, A., and Solomatine, D. (1998). "Application of Global Optimization to the Design of Pipe Networks." *Proc. 3rd Intern Conf. Hydroinformatics, Copenhagen*, 989-996.

Abbott, M. B. (1979). *Computational hydraulics : elements of the theory of free surface flows / M.B. Abbott*. Monographs and surveys in water resources engineering ; 1, Pitman Pub. ; Fearon-Pitman Publishers, London : Belmont, Calif. :

Abbott, M., and Cunge, J. (1986). *Engineering Applications of Computational Hydraulics: Elements of the Theory of Free Surface Flows v. 1*. Addison-Wesley Educational Publishers Inc.

Abbott, M. B. (1991). *Hydroinformatics: Information Technology and the Aquatic Environment*. Avebury.

Alfieri, L., Laio, F., and Claps, P. (2008). "A simulation experiment for optimal design hyetograph selection." *Hydrological Processes*, 22(6), 813-820.

Allen, R. J., and DeGaetano, A. T. (2005). "Considerations for the use of radar-derived precipitation estimates in determining return intervals for extreme areal precipitation amounts." *Journal of Hydrology*, 315(1-4), 203-219.

Amdahl, G. (1967). "Validity of the single processor approach to achieving large scale computing capabilities." *Proceedings of the April 18-20, 1967, spring joint computer conference*, ACM New York, NY, USA, Atlantic City, New Jersey, 483-485.

Ana, E. J., and Bauwens, W. (2007). "Review of Sewer Asset Management Decision Support Systems." UNESCO, Paris.

Anderson, D. P., Cobb, J., Korpela, E., Lebofsky, M., and Werthimer, D. (2002). "SETI@home: an experiment in public-resource computing." *Commun. ACM*, 45(11), 56-61.

Baik, H., Jeong, H. S. (., and Abraham, D. M. (2006). "Estimating Transition Probabilities in Markov Chain-Based Deterioration Models for Management of Wastewater Systems." *Journal of Water Resources Planning and Management*, 132(1), 15-24.

Barraud, S., Azzout, Y., Cres, F., and Chocat, B. (1999). "Selection aid of alternative techniques in urban storm drainage - proposition of an expert system." <http://www.iwaponline.com/wst/03904/wst039040241.htm> (Feb. 4, 2010).

Barreto, W., Vojinovic, Z., Price, R., and Solomatine, D. P. (2006). "Approaches to Multi-objective multi-tier Optimization in Urban Drainage Planning." *Proc. 7th Intern. Conf. on Hydroinformatics*, Nice, France.

Beowulf.org: The Beowulf Cluster Site. (2010). <http://www.beowulf.org/> (Feb. 15, 2010).

Blazewicz, J., Ecker, K. H., and Yang, T. (2000). "New trends on scheduling in parallel and distributed systems." *Parallel Computing*, 26(9), 1061-1063.

Broad, D., Dandy, G., and Maier, H. (2005). "Water Distribution System Optimization Using Metamodels." *Journal of Water Resources Planning and Management*, 131(3), 172-180.

Butler, D., and Davies, J. W. (2004). *Urban drainage*. Taylor & Francis.

CERN openlab Phase I - opencluster: Overview. (2010). <https://openlab-mu-internal.web.cern.ch/openlab-mu-internal/10_openlab-I/opencluster/default.asp> (Feb. 15, 2010).

Calver, A. (1993). "THE TIME-AREA RUNOFF FORMULATION REVISITED.." *Proceedings of the ICE - Water Maritime and Energy*, 101(1), 31-36.

Černý, V. (1985). "Thermodynamical approach to the traveling salesman problem: An efficient simulation algorithm." *Journal of Optimization Theory and Applications*, 45(1), 41-51.

Chow, V., Maidment, D., and Mays, L. (1988). *Applied Hydrology*. McGraw-Hill Science/Engineering/Math.

Chapman, B., Jost, G., and Pas, R. V. D. (2007). *Using OpenMP: Portable Shared Memory Parallel Programming*. The MIT Press.

Clemens, F. (1998). "Hydrodynamic Models in Urban Drainage: Application and Calibration." DUP Science - Delft University Press.

Colyer, P. (1977). "PERFORMANCE OF STORM DRAINAGE SIMULATION MODELS.." *ICE Proceedings*, 63(2), 293-309.

Coello, C. A. C., Veldhuizen, D. A. V., and Lamont, G. B. (2002). *Evolutionary Algorithms for Solving Multi-objective Problems*. Springer.

Collette, Y., and Siarry, P. (2003). *Multiobjective optimization*. Springer.

Deb, K., Pratap, A., Agarwal, S., and Meyarivan, T. (2002). "A fast and elitist multiobjective genetic algorithm : NSGA-II." *Evolutionary Computation, IEEE Transactions on*, 6(2), 197, 182.

Djordjevic, S., Prodanovic, D., and Walters, G. A. (2004). "Simulation of Transcritical Flow in Pipe/Channel Networks." *Journal of Hydraulic Engineering*, 130(12), 1167-1178.

Dorigo, M., Maniezzo, V., and Colorni, A. (1996). "The Ant System: Optimization by a colony of cooperating agents." *IEEE Transactions on Systems, Man, and*

Cybernetics Part B: Cybernetics, 26(1), 41, 29.

Eberhart, R., and Kennedy, J. (1995). "A new optimizer using particle swarm theory." *Micro Machine and Human Science, 1995. MHS '95., Proceedings of the Sixth International Symposium on*, 43, 39.

Elleuch, A., Kanawati, R., Muntean, T., and Talbi, E. (1994). "Dynamic load balancing mechanisms for a parallel operating system kernel." *Parallel Processing: CONPAR 94 — VAPP VI*, 866-877.

Fang, Z., Haymet, A., Shinoda, W., and Okazaki, S. (1999). "Parallel molecular dynamics simulation: Implementation of PVM for a lipid membrane." *Computer Physics Communications*, 116(2-3), 295-310.

Geist, A. (2005). "New Directions in PVM/Harness Research." *Recent Advances in Parallel Virtual Machine and Message Passing Interface*, 1-3.

Geist, A., Beguelin, A., Dongarra, J., Jiang, W., Manchek, R., and Sunderam, V. S. (1994). *PVM: Parallel Virtual Machine: A Users' Guide and Tutorial for Network Parallel Computing*. The MIT Press.

Geist, G. A., Kohl, J. A., and Papadopoulos, P. M. (1996). "PVM and MPI: a Comparison of Features." *CALCULATEURS PARALLELES*, 8, 137—150.

Geetha, K., Mishra, S. K., Eldho, T. I., Rastogi, A. K., and Pandey, R. P. (2007). "Modifications to SCS-CN Method for Long-Term Hydrologic Simulation." *Journal of Irrigation and Drainage Engineering*, 133(5), 475-486.

Gill, T. D. (2005). "Transformation of Ppoint Rainfall to Areal Rainfall by Estimating Areal Reduction Factors, Using Radar Data, Fortexas." Texas A&M University.

Goel, T., Vaidyanathan, R., Haftka, R. T., Shyy, W., Queipo, N. V., and Tucker, K. (2007). "Response surface approximation of Pareto optimal front in multi-objective optimization." *Computer Methods in Applied Mechanics and Engineering*, 196(4-6), 879-893.

Goldberg, D. E. (1989). *Genetic Algorithms in Search, Optimization, and Machine Learning*. Addison-Wesley Professional.

Grimaldi, S., and Serinaldi, F. (2006). "Design hyetograph analysis with 3-copula function / Analyse de hyétogrammes de projet avec une fonction 3-copules." *Hydrological Sciences Journal*, 51(2), 223-338.

Gropp, W., Lusk, E., and Skjellum, A. (1994). *Using MPI: Portable Parallel Programming with the Message-Passing Interface*. The MIT Press.

Gropp, W., and Lusk,Ewing. (2002). "Goals Guiding Design: PVM and MPI." *Fourth IEEE International Conference on Cluster Computing (CLUSTER'02)*, Chicago, Illinois, 257-265.

Gustafson, J. L. (1988). "Reevaluating Amdahl's Law." *COMMUNICATIONS OF THE ACM*, 31, 532—533.

Hahn, M. A., Palmer, R. N., Merrill, M. S., and Lukas, A. B. (2002). "Expert System for Prioritizing the Inspection of Sewers: Knowledge Base Formulation and Evaluation." *Journal of Water Resources Planning and Management*, 128(2), 121-129.

Hajela, P., and Lin, C. -. (1992). "Genetic search strategies in multicriterion optimal design." *Structural and Multidisciplinary Optimization*, 4(2), 99-107.

Hawkins, R. H. (1978). "Runoff curve number with varying site moisture." *J. Irrig. and Drain. Div.*, 104(4), 389-398.

Hillier, F. S., and Lieberman, G. J. (2002). *Introduction to Operations Research*. McGraw-Hill Science/Engineering/Math.

Hluchý, L., Tran, V., Halada, L., and Dobrucký, M. (1999). "Ground Water Flow Modelling in PVM." *Recent Advances in Parallel Virtual Machine and Message Passing Interface*, 70.

Holland, J. H. (1992). *Adaptation in Natural and Artificial Systems: An Introductory Analysis with Applications to Biology, Control, and Artificial Intelligence*. The MIT Press.

Huff, F. A. (1967). "Time Distribution of Rainfall in Heavy Storms." *Water Resources Research*, 3(4), 1007-1019.

Hughes, C., and Hughes, T. (2003). *Parallel and Distributed Programming Using C++*. Addison-Wesley Professional.

Izquierdo, J., Montalvo, I., Pérez, R., and Fuertes, V. S. (2008). "Design optimization of wastewater collection networks by PSO." *Comput. Math. Appl.*, 56(3), 777-784.

Jarboui, B., Cheikh, M., Siarry, P., and Rebai, A. (2007). "Combinatorial particle swarm optimization (CPSO) for partitional clustering problem." *Applied Mathematics and Computation*, 192(2), 337-345.

Jin, Y., Okabe, T., and Sendho, B. (2001). "Adapting Weighted Aggregation for Multiobjective Evolution Strategies." *Evolutionary Multi-Criterion Optimization*, 96-110.

Jin, S., Schiavone, G., and Turgut, D. (2008). "A performance study of multiprocessor task scheduling algorithms." *J. Supercomput.*, 43(1), 77-97.

Jiang, C., Li, K., Liu, N., and Zhang, Q. (2005). "Implicit Parallel FEM Analysis of Shallow Water Equations." *Tsinghua Science & Technology*, 10(3), 364-371.

Karniadakis, G., and Kirby, R. (2003). *Parallel Scientific Computing in C++ and MPI: A Seamless Approach to Parallel Algorithms and their Implementation*. Cambridge

University Press.

Keifer, C., and Chu, H. (1957). "Synthetic storm pattern for drainage design." *Journal of Hydraulic Procedures*, 83(4), 1334-1352.

Kirkpatrick, S., Gelatt, C. D., and Vecchi, M. P. (1983). "Optimization by Simulated Annealing." *Science*, 220(4598), 671-680.

Knowles, J. (2006). "ParEGO: a hybrid algorithm with on-line landscape approximation for expensive multiobjective optimization problems." *IEEE Transactions on Evolutionary Computation*, 10(1), 50-66.

Knowles, J., and Hughes, E. J. (2005). "Multiobjective Optimization on a Budget of 250 Evaluations." *Evolutionary Multi-Criterion Optimization*, 176-190.

Kutija, V. (1993). "On the numerical modelling of supercritical flow." *Journal of Hydraulic Research*, 31(6), 841.

Laumanns, M., Thiele, L., Deb, K., and Zitzler, E. (2002). "Combining convergence and diversity in evolutionary multiobjective optimization." *Evol. Comput.*, 10(3), 263-282.

Lekuthai, A., and Vongvisessomjai, S. (2001). "Intangible Flood Damage Quantification." *Water Resources Management*, 15(5), 343-362.

Lin, G., Wu, M., Chen, G., and Liu, S. (2009). "Construction of design hyetographs for locations without observed data." *Hydrological Processes*.

Lyn, D. A., and Altinakar, M. (2002). "St. Venant--Exner Equations for Near-Critical and Transcritical Flows." *Journal of Hydraulic Engineering*, 128(6), 579-587.

Martin, C., Ruperd, Y., and Legret, M. (2007). "Urban stormwater drainage management: The development of a multicriteria decision aid approach for best management practices." *European Journal of Operational Research*, 181(1), 338-349.

Maniezzo, V., and Roffilli, M. (2008). "Very Strongly Constrained Problems: an Ant Colony Optimization Approach." *Cybernetics and Systems*, 39, 395-424.

Marsalek, J. (1984). "Head Losses at Sewer Junction Manholes." *Journal of Hydraulic Engineering*, 110(8), 1150-1154.

Mays, L. W. (2001). *Stormwater collection systems design handbook*. McGraw-Hill Professional.

McKay, M. D., Beckman, R. J., and Conover, W. J. (1979). "A Comparison of Three Methods for Selecting Values of Input Variables in the Analysis of Output from a Computer Code." *Technometrics*, 21(2), 239-245.Mahinthakumar, G., and Saied, F. (2002). "A Hybrid MPI-OpenMP Implementation of an Implicit Finite-Element Code on Parallel Architectures." *International Journal of High Performance Computing Applications*, 16(4), 371-393.

Meselhe, E. A., and Holly Jr., F. M. (1997). "Invalidity of Preissmann Scheme for Transcritical Flow." *Journal of Hydraulic Engineering*, 123(7), 652-655.

Meselhe, E. A., and Holly, J. (1993). "Simulation of Unsteady Flow in Irrigation Canals with Dry Bed." *Journal of Hydraulic Engineering*, 119(9), 1021-1039.

Mishra, S. K., and Singh, V. P. (2003). *Soil conservation service curve number (SCS-CN) methodology*. Springer.

Moore, M. (2004). "An accurate parallel genetic algorithm to schedule tasks on a cluster." *Parallel Comput.*, 30(5-6), 567-583.

Myerson, J. (2009). "Cloud computing versus grid computing." CT316, , <http://www.ibm.com/developerworks/web/library/wa-cloudgrid/> (Feb. 15, 2010).

Nascimento, N. D. O., Balbi, D. A. F., and Naghettini, M. (2000). "Modeling the Time Distributions of Heavy Storms --- Design Hyetographs." *ASCE Conf. Proc.*, ASCE, Minneapolis, Minnesota, USA, 35.

Nelder, J. A., and Mead, R. (1965). "A Simplex Method for Function Minimization." *The Computer Journal*, 7(4), 308-313.

NRCS. (1986). *Urban Hydrology for Small Watersheds. TR-55*. United States Department of Agriculture.

Omolayo, A. (1993). "On the transposition of areal reduction factors for rainfall frequency estimation." *Journal of Hydrology*, 145(1-2), 191-205.

Osyczka, A. (1984). *Multicriterion Optimization in Engineering with Fortran Programmes*. Ellis Horwood Ltd , Publisher.

Ouenes, A., Weiss, W., Sultan, A., and Anwar, J. (1995). "Parallel Reservoir Automatic History Matching Using a Network of Workstations and PVM." *Proceedings of SPE Reservoir Simulation Symposium*.

Pongchairerks, P., and Kachitvichyanukul, V. (2009). "A Particle Swarm Optimization Algorithm on Job-Shop Scheduling Problems with Multi-Purpose Machines." *APJOR*, 26(2), 184, 161.

Prékopa, A. (1995). *Stochastic Programming*. Springer.

Prodanovic, P., and Simonovic, S. (2004). "Assessment of water resources risk and vulnerability to changing climatic conditions. Report V:Generation of synthetic design storms for the Upper Thames River basin." The Canadian Foundation for Climate and Atmospheric Sciences.

Rodi, W. (1993). *Turbulence models and their application in hydraulics*. Taylor & Francis.

Rodriguez-Iturbe, I., and Mejía, J. M. (1974). "On the Transformation of Point Rainfall to Areal Rainfall." *Water Resources Research*, 10(4), PAGES 729–735.

Saegrov, S. (2006). *CARE-S*. IWA Publishing.

Saghafian, B., and Bondarabadi, S. R. (2008). "Validity of Regional Rainfall Spatial Distribution Methods in Mountainous Areas." *Journal of Hydrologic Engineering*, 13(7), 531-540.

Serrano, S. E. (2003). "Improved Decomposition Solution to Green and Ampt Equation." *Journal of Hydrologic Engineering*, 8(3), 158-160.

Shinji, A., and Tetsuya, K. (2005). "Head Loss at Three-Way Circular Drop Manhole." *Journal of Japan Sewage Works Association*, 42(510), 124-132.

Sivapalan, M., and Blöschl, G. (1998). "Transformation of point rainfall to areal rainfall: Intensity-duration-frequency curves." *Journal of Hydrology*, 204(1-4), 150-167.

Sloan, J. D. (2004). *High Performance Linux Clusters with OSCAR, Rocks, OpenMosix, and MPI*. O'Reilly Media.

Solomatine, D. P. (1999). "Two Strategies of Adaptive Cluster Covering with Descent and Their Comparison to Other Algorithms." *J. of Global Optimization*, 14(1), 55-78.

Solomatine, D. P. (. (2005). "Adaptive Cluster Covering and Evolutionary Approach: Comparison, Differences and Similarities." *Proc. IEEE Congress on Evolutionary Computation*, 1959-1966.

Stone, S., Dzuray, E., Meisegeier, D., Dahlborg, A., and Ericson, M. (2002). *Decision-Support Tools for Predicting the Performance of Water Distribution and Wastewater Collection Systems*. U.S. Environmental Protection Agency, 101.

Sudarsan, R., and Ribbens, C. J. (2010). "Design and performance of a scheduling framework for resizable parallel applications." *Parallel Computing*, 36(1), 48-64.

Tanaka, N., and Tatano, H. (2000). "A performance evaluation method of monitoring systems for inner-basin drainage under imperfect observation." *Hydrological Processes*, 14(3), 621-638.

Velte, T., Velte, A., and Elsenpeter, R. (2009). *Cloud Computing, A Practical Approach*. McGraw-Hill Osborne Media.

Vamvakeridou-Lyroudia, L. S., Walters, G. A., and Savic, D. A. (2005). "Fuzzy Multiobjective Optimization of Water Distribution Networks." *Journal of Water Resources Planning and Management*, 131(6), 467-476.

Vojinovic, Z., Solomatine, D., and Price, R. K. (2006). "Dynamic least-cost optimisation of wastewater system remedial works requirements." *Water Science and Technology: A Journal of the International Association on Water Pollution Research*, 54(6-7), 467-475.

Vojinovic, Z., Price, R., and Den Broek, W. (2005). "HYDROPLAN-EU Knowledge Management Framework for Urban Water Asset Management." Copenhagen.

Watkins, L. (1976). "THE TRRL HYDROGRAPH METHOD OF URBAN SEWER DESIGN ADAPTED FOR TROPICAL CONDITIONS.." *ICE Proceedings*, 61(3), 539-566.

Watt, W., Chow, K., Hogg, W., and Lathem, K. (1986). "A 1-h urban design storm for Canada." *Canadian Journal of Civil Engineering*, 13(3), 293-300.

Witter, J. V. (1983). "StatisticalAreal Reduction Factor in The Netherlands." *Proc. Hamburgo Symposium*, IAHS Publ., 25-34.

Wulf, W., Cohen, E., Corwin, W., Jones, A., Levin, R., Pierson, C., and Pollack, F. (1974). "HYDRA: the kernel of a multiprocessor operating system." *Commun. ACM*, 17(6), 337-345.

Yen, B. C., and Chow, V. T. (1980). "Design hyetographs for small drainage structures." *Journal of the Hydraulics Division*, 106(6), 1055-1076.

Yen, B. C., and Pansic, N. (1980). "Surcharge of Sewer Systems." WRC Reseach Report No 149.

Zhao, C., Zhu, D. Z., and Rajaratnam, N. (2008). "Computational and Experimental Study of Surcharged Flow at a 90[degree] Combining Sewer Junction." *Journal of Hydraulic Engineering*, 134(6), 688-700.

Zhou, F., Hicks, F., and Steffler, P. (2004). "Analysis of effects of air pocket on hydraulic failure of urban drainage infrastructure." *Canadian Journal of Civil Engineering*, 31(1), 86-94.

About the Author

Wilmer Jóse Barreto Cordero was born in Barquisimeto,Venezuela on December 11th, 1966. In 1992, he graduated as Civil Engineer at "Lisandro Alavarado" university in Barquisimeto. He was always attracted by computers and programming which took him to develop several informatics tools during his career, finalizing his graduation with the implementation of daily rainfall-runoff model for his degree thesis. After his graduation, he moved to Mérida, working as assistant engineer during two year at The Inter-American Center for development of Land and Water (CIDIAT). During that time he was in charge of setting up models for the institute, he deals with hydrological models as HPSF and SIHIM, and with the modeling of ground-water using GW and MODFLOW. He participates in several projects in Venezuela and abroad.

In 1994, he returned to Barquisimeto as lecturer at the "Lisandro Alvarado" University in the area of hydraulic and fluid mechanics. In 1995, he met his former hydraulic lecturer Dr. Nestor Méndez, who was coming back from IHE after finalize his PhD at Wageningen University. Dr Méndez encouraged him to study Hydroinformatics at IHE, which was starting to become of importance among hydraulic engineers. In 1999, he is awarded with a scholarship from the World Bank (WB/JJP) to study at IHE in Delft, and in 2001 he got the Master of Science degree in Hydroinformatics. He returned to his former university as lecturer and worked in consultancy until he got a scholarship to do a PhD at TU-Delft and Unesco-IHE. During that time he returned in 2010 to Venezuela and worked as lecturer and consultant while tried to finalize the thesis writing, it was finally done year and half after his arrival to Venezuela.

List of Publications

Barreto, W., Vojinovic, Z., Price, R., and Solomatine, D. (2010). "Multiobjective Evolutionary Approach to Rehabilitation of Urban Drainage Systems." *Journal of Water Resources Planning and Management*, 136(5), 547.

Barreto, W., Vojinovic, Z., Price, R. K., and Solomatine, D. P. (2006). "Approaches to Multi-Objective Multi-Tier Optimization in Urban Drainage Planning." *7th International Conference on Hydroinformatics*, Nice, FRANCE.

Barreto, W., Vojinovic, Z., Price, R. K., and Solomatine, D. P. (2008). "Multi-tier Modelling of Urban Drainage Systems: Muti-objective Optimization and Parallel Computing." *11th International Conference on Urban Drainage*, Edinburgh, Scotland, UK.

Corzo, G. A., Barreto, W., Love, D., and Uhlenbrook, S. (2010). "HBVX: A Semi-Distributed Hydrological Modelling Tool With Multi-Criteria Calibration." *9th International Conference on Hydroinformatics*, Tianjin, CHINA, 1982-1990.

Vojinovic, Z., Sánchez, A., and Barreto, W. (2008). "Optimising Sewer System Rehabilitation Strategies between Flooding , Overflow Emissions and Investment Costs." *11th International Conference on Urban Drainage*, Edinburgh, Scotland, UK.

Samenvatting

Overstromingen in stedelijke gebieden is wereldwijd een ernstig probleem geworden. Het prestatieniveau van stedelijk water afvoer ('urban drainage systems' (UDS) degradeert om een aantal redenen: structurele verslechtering, aanslibbing, nieuwe ontwikkelingen, klimaatverandering etc. Om een acceptabele prestatie van de UDS te behouden, dienen vroegtijdig onderhoudsplannen ontwikkeld en geïmplementeerd te worden.

Steden groeien snel, en het budget voor het onderhoud van draineringstelsels groeit in een veel trager tempo dan die voor stedelijke ontwikkeling. In ontwikkelingslanden is de situatie ernstig, er wordt weinig geïnvesteerd en er zijn jaarlijks minder fondsen voor het onderhoud beschikbaar De toewijzing van zulke onderhoudsfondsen dient 'optimaal' te zijn om waar voor het geld te krijgen. Deze taak is echter niet makkelijk te realiseren vanwege de vele diverse criteria voor het onderhoudsproces, rekening houdend met belangen op het technische, ecologische en sociale vlak. Vaak zijn deze criteria in strijd met elkaar, waardoor het een veeleisende taak wordt.

Multi-objective optimalisatie benaderingen nemen passende programma's en faciliteiten aan om het optimale herstel van een UDS te simplificeren. Heuristische en Genetische algoritmes zijn toegepast en bewijzen efficiënt te zijn voor multi-objective problemen. Echter, het grote aantal mogelijke oplossingen (of scenario's) in UDS en het aantal die nodig zijn voor bijvoorbeeld evolutionaire algoritmen (EA) bemoeilijkt de toepassing ervan voor beoefenaars.

Het huidige onderzoek heeft als doel een kader te definiëren om met multi-criteria besluitvorming over het onderhoud van rioleringsstelsels (stedelijke afvoersystemen) te handelen, en richt zich op een aantal aspecten, zoals prestatieverbetering van de multi-criteria optimalisatie door middel van het opnemen van nieuwe kenmerken in de algoritmes en de juiste selectie van prestatiecriteria. Het nieuwe kader, 'Multi-tier Approach' genoemd, dient geschikt te zijn om in ontwikkelingslanden toe te passen, voor diverse schaalgroottes en dient verschillende oplossingen te kunnen bieden in een bepaalde tijd die aanvaardbaar is voor de gebruikers.

Er is een overzicht gemaakt van de stand van zaken van het onderhoud van de stedelijke riolering. Tijdens het onderhoudsproces dienen verscheidene aspecten te worden aangepakt. Er dient rekening gehouden te worden met zaken zoals het vaststellen van de prestatie-indicatoren voor hydraulische, structurele en milieu toetsing. De beschikbaarheid van data en de identificatie van kritische pijpen en tunnels zijn tevens erg belangrijk in een onderhoudsplan. Hydrologische en hydrodynamische modellering vormen een sleutelrol tijdens de hydraulische, structurele en milieu toetsing. Een koppeling van de boven- en ondergrondse systemen is noodzakelijk om de consequenties van mogelijke overloop te kunnen evalueren Momenteel bestaan er volledige geautomatiseerde 1D en 2D modelleringpakketten, echter, de interactie tussen deze modellen, veroorzaakt door deze modellen vergt nog onderzoek om beschikbaar te worden voor de gebruikers. Een duurzame benadering, gericht op de controle van de hoeveelheid afvoer vanaf het begin van de neerslag, heeft de voorkeur boven methodes gebaseerd op overdracht. Deze duurzame benadering is tevens gericht op het onderhouden van een balans tussen ecologische, sociale en economische waarden. Om voor dit onderzoek de meest geschikte pakketten te selecteren, zijn verschillende hydrodynamische modelleringpakketten, gebaseerd op hun voor- en nadelen, herzien.

Er is een benadering ontwikkeld voor rioleringsonderhoud welke gebruik maakt van een multi-tier kader. Deze benadering is innovatief omdat het hydrodynamische modellering introduceert binnen een multi-objectieve optimalisatie proces, waarbij gebruik gemaakt wordt van parallel computergebruik. Hierdoor wordt het gebruik aantrekkelijker voor de gebruiker. Afhankelijk van de beschikbaarheid van data is het modulair en flexibel, waardoor het geschikt is om in ontwikkelingslanden te gebruiken. Kennis kan in verschillende stadia van het optimisatieproces opgenomen worden. De kaderapplicatie is simpel maar laat geen belangrijke kenmerken voor de rioleringsonderhoudsystemen achterwege. Het gebruik van externe onderdelen voor modellering, optimalisatie en visualisatie staan schaalbaarheid toe, wat impliceert dat de onderdelen naar gelang kunnen groeien. Een prototype kader is ontworpen en met succes getest op reële problemen in ontwikkelingslanden.

Een overzicht van de huidige stand van zaken van de optimimalisatie van bouwkundige problemen is afgerond. Het laat zien dat er niet een unieke methode is om verschillende soorten problemen te optimaliseren, in plaats daarvan dient een geschikte methodiek gekozen te worden die rekening houdt met de problemen die geoptimaliseerd dienen te worden. Er zijn weinig applicaties gevonden op het gebied van stedelijke riolering, en geen enkele applicatie, mits gebruikt op kleinschalige netwerken, voor reële gevallen waarbij gebruik gemaakt wordt van hydrodynamische modellen binnen de optimalisatie algoritme. Een belangrijk probleem betreft de benodigde rekencapaciteit.. In sommige literatuur wordt dit een 'duur probleem' genoemd. Multi-criteria rioleringsstelselonderhoud wordt gezien als een probleem waar veel automatiseringskracht voor nodig is. Een benadering om dit probleem het hoofd te bieden is ontwikkeld en getest voor multi-objectieve algoritmen waarbij gebruik gemaakt is van vier criteria functies. De methode is gebaseerd op de veronderstelling dat gebruikers maar enkele oplossingen nodig hebben en niet een hele reeks die op elkaar lijken. De nieuwe benadering presteert op drie van deze criteria beter dan NSGA-II, terwijl NSGA-II iets beter presteerde op het vierde criterium.

Het voorgestelde multi-tier kader voor het herstel van een stedelijk rioleringssysteem is geïmplementeerd en getest, waarbij het principe is aangetoond op een kleinschalige studie. Ten eerste is er een constructie geïmplementeerd om de investeringskosten te schatten; functies als de vervanging van pijpen, opslagtanks en –plassen en afwijkende constructies zijn meegenomen. Een methode voor de schatting van schadekosten is tevens opgenomen. De kostenconstructie is gebaseerd op vergelijkingen in gemeenschappelijk gebruik en volgen een uniforme prijs analyse, inclusief de vooraf ingestelde waarde in de kostenschatting.

Een prototype bewerking is ontwikkeld en getest. Een kleinschalige studie is gebruikt om het principe aan te tonen Twee multicriteria algoritmen gebaseerd op genetische algoritmen zijn getest: dit waren NSGA-II en ε-MOEA. NSGA-II is opgenomen in een bibliotheek en in een optimalisatie pakket genaamd NSGAX Om de prestaties van de algoritmes te kunnen vergelijken, zijn vier metrische indicatoren van multi-criteria uitvoeringen gebruikt: cardinaliteit, tijd, aantal evaluatie functies, hypervolume en ε-Indicator. Door het gebruik van deze indicatoren is geconcludeerd dat NSGA-II over het algemeen ε-MOEA beter presteert, maar de verschillen waren niet substantieel. . Hypervolume en ε-Indicator laten convergentie zien, en zij werden gebruikt als stop criteria tijdens het optimalisatieproces. De toepassing van het kader op een groter netwerk van 63 pijpen liet de behoefte aan meer efficiënte algoritmen of computers zien om de rekentijd te verminderen. In beide studies van 12 en 63 pijpen was het mogelijk om rationele oplossingen te vinden die door uitvoerders verwacht konden worden; bijvoorbeeld de opslagtanks waren correct geselecteerd afhankelijk van de locatie en zij waren ook naar behoren afgemeten. Geconcludeerd kan worden dat het ontwikkelde kader geschikt is voor gebruik in het herstel van stedelijke rioleringssystemen. Het laat schaalbaarheid en flexibiliteit zien. Het gebruik van immateriële kosten is tevens geëvalueerd door middel van de implementatie van een objectieve functie om de ongerustheid van de bevolking te minimaliseren. Het programma laat interpretatie en onderhandeling over verschillende alternatieven toe tussen de belanghebbenden en is niet afhankelijk van één optimale oplossing.

De stand van zaken betreffende parallellisme voor computer codes is herzien. Voordelen en nadelen van de verschillende bestaande methodes voor parallelle applicatie zijn bestudeerd. De theorie van simultaneïteit voor het gebruik van meerder computer processors en clusters is herzien. Er is een parallelle code voor NSGAX geïmplementeerd door het gebruik van PVM bibliotheken over Cygwin, een Linux OS emulator voor Windows. De parallelle code is in het mutli-tier optimalisatie kader opgenomen en is op twee studies toegepast.

Er is een kleinschalig cluster opgezet bestaande uit heterogene PC's met enkelvoudige en multi-core processors. Twee onderzoeken zijn getest op het parallelle kader: één gebruikmakend van een kleinschalig netwerk met 12 pijpen en de ander voor een substroomgebied in Belo Horizonte Brazilië, bestaande uit 168 pijpen. De resultaten laten een goede besparing van de verbruikte tijd zien tussen 60% en 80% in vergelijking met de prestatie van enkelvoudige optimalisatie. Tevens kan geconcludeerd worden dat het aantal, dat gebruikt dienen te worden in een cluster gerelateerd dienen te zijn aan de grootte van het probleem; als de resultaten geanalyseerd worden door middel van een efficiency indicator, is het beter om minder processors te gebruiken. Wanneer het probleem met 12 pijpen opgelost werd met zes processors gedroegen deze zich als alleen drie processors in termen van het versnellen, terwijl de zes processors van het Bel Horizonte onderzoek zich als vijf gedroegen. Het is dus efficiënter voor een groter dan voor een klein probleem.

Toepassingen van stedelijk rioleringsherstel met multi-objective algoritmen gebaseerd op bevolkingsoptimalisatie zoals GA is gelimiteerd onder de gebruikers. Bovendien zijn er minder gebruikers die hydrodynamische modellen toepassen om de overstromingschade te schatten. Dit komt omdat de gebruikte methodieken rekenkundig veeleisend zijn, waardoor gebruikers aangemoedigd worden om deze toepassingen te negeren. De ontwikkelde multi-tier benadering is toegepast op twee uitgevoerde onderzoeken in ontwikkelingslanden. Eén onderzoek was in Cabudare, Venezuela en het andere in Cali, Colombia. Aangetoond is dat een multi-tier kader tevens toepasbaar is in ontwikkelingslanden waar data gelimiteerd beschikbaar is en waar gesimplificeerde programma's erkend zijn als waardevol voor gebruikers.

In de Cabudare studie is een benadering toegepast waarbij gebruik is gemaakt van verwachtte schadekosten. Dit heeft geresulteerd in drie gelijke economische oplossingen betreffende de Pareto reeks. Een vergelijking met de aanbevolen benadering van de Venezolaanse standaard laat zien dat deze drie oplossingen gelijkwaardig zijn aan de vaststelling van de complete overstromingsschadeposten gedurende een herhalingstijd van neerslag van 1:10. Het biedt ook een aantal verschillende oplossingen waarin er een afweging tussen investeringen en schade wordt gemaakt.

De NSGAX algoritme is verbeterd om de deskundige vakkennis in te kunnen voegen, door middel van een injectie van goede genen in de oorspronkelijke populatie. Deze nieuwe benadering heeft bewezen efficiënt te zijn voor veeleisende computationele problemen.

In de Cali studie is de milieu variabele aangepakt. De multi-tier kader is gekoppeld aan SWMM 5.0 om de overstroming en waterkwaliteit te berekenen. Een afwijking is aangebracht om water van het netwerk naar een tijdelijke opslag te leiden om zo het overschot te reduceren en om het eerste overtollige water op te vangen. Ondanks de simplificatie van het opslagmodel, het nadeel tussen de investeringskosten en de overstromingsschade, tussen de investeringskosten en de waterkwaliteit en de correlatie tussen overstromingsschade en de waterkwaliteit, is er verbetering te zien.

Het doel van het ontwikkelen en testen van een model voor het herstel van rioleringsstelsels is daarom behaald. Een prototype van de 'multi-tier' benadering is toegepast op verschillende teststudies in ontwikkelingslanden, die laat zien dat het makkelijk en toepasbaar is. Parallel computergebruik en andere methodieken maken deze benadering aantrekkelijk voor uitvoerders, en deze methodieken zijn in het model verwerkt.

Flooding in urbanized areas has become a very important issue around the world. The level of service (or performance) of urban drainage systems (UDS) degrades in time for a number of reasons. In order to maintain an acceptable performance of UDS, early rehabilitation plans must be developed and implemented. In developing countries the situation is serious, little investment is done and there are smaller funds each year for rehabilitation. The allocation of such funds must be 'optimal' in providing value for money. However this task is not easy to achieve due to the multicriteria nature of the rehabilitation process, taking into account technical, environmental and social interests. Most of the time these are conflicting, which make it a highly demanding task.

The present book introduce a framework to deal with multicriteria decision making for the rehabilitation of urban drainage systems, and focuses on several aspects such as the improvement of the performance of the multicriteria optimization through the inclusion of new features in the algorithms and the proper selection of performance criteria. The use of Genetic Algorithms, parallelization and application in countries like Brazil, Colombia y Venezuela are treated in this book.

CRC Press
Taylor & Francis Group

A BALKEMA BOOK

ISBN-13: 978-0-415-62478-7

9 780415 624787

an **informa** business

T - #0223 - 071024 - C0 - 240/170/14 - PB - 9780415624787 - Gloss Lamination